Quick Guide

WIRING

CREATIVE HOMEOWNER PRESS®

COPYRIGHT © 1992
CREATIVE HOMEOWNER PRESS®
A Division of Federal Marketing Corp.
Upper Saddle River, NJ

Editor: Warren Ramezzana
Project Editor: Kimberly Kerrigone
Senior Designer: Annie Jeon
Illustrators: James Randolph, Norman Nuding
Production Assistant: Mindy Circelli
Technical Reviewer: Jim Barrett

Cover Design: Warren Ramezzana
Cover Illustrations: Moffit Cecil

Electronic Prepress: M. E. Aslett Corporation
Printed at: Banta Company

Current Printing (last digit)
10 9 8 7 6 5 4 3

Quick Guide: Wiring
LC: 92-81623
ISBN: 1-880029-13-8 (paper)

CREATIVE HOMEOWNER PRESS®
A Division of Federal Marketing Corp.
24 Park Way
Upper Saddle River, NJ 07458

CONTENTS

SAFETY FIRST

Though all the designs and methods in this book have been tested for safety, it is not possible to overstate the importance of using the safest construction methods possible. What follows are reminders; some do's and don'ts of basic carpentry. They are not substitutes for your own common sense.

- *Always* use caution, care, and good judgment when following the procedures described in this book.

- *Always* turn off the power at the main electrical service panel before beginning work.

- *Always* be sure that the electrical setup is safe; be sure that no circuit is overloaded, and that all power tools and electrical outlets are properly grounded. Do not use power tools in wet locations.

- *Always* use tools that have insulated handles. Do not use a metal ladder when working with electricity.

- *Always* read the tool manufacturer's instructions for using a tool, especially the warnings.

- *Always* use holders or pushers to work pieces shorter than 3 inches on a table saw or jointer. Avoid working short pieces if you can.

- *Always* remove the key from any drill chuck (portable or press) before starting the drill.

- *Always* know the limitations of your tools. Do not try to force them to do what they were not designed to do.

- *Always* make sure that any adjustment is locked before proceeding. For example, always check the rip fence on a table saw or the bevel adjustment on a portable saw before starting to work.

- *Always* clamp small pieces firmly to a bench or other work surfaces when sawing or drilling.

- *Always* wear the appropriate rubber or work gloves when handling chemicals, heavy construction or when sanding.

- *Always* wear a disposable mask when working with odors, dusts or mists. Use a special respirator when working with toxic substances.

- *Always* wear eye protection, especially when using power tools or striking metal on metal or concrete; a chip can fly off, for example, when chiseling concrete.

- *Always* be aware that there is never time for your body's reflexes to save you from injury from a power tool in a dangerous situation; everything happens too fast. Be *alert!*

- *Always* keep your hands away from the business ends of blades, cutters and bits.

- *Always* hold a portable circular saw with both hands so that you will know where your hands are.

- *Always* use a drill with an auxiliary handle to control the torque when large size bits are used.

- *Always* check your local building codes when planning new construction. The codes are intended to protect public safety and should be observed to the letter.

- *Never* work with power tools when you are tired or under the influence of alcohol or drugs.

- *Never* cut very small pieces of wood or pipe. Whenever possible, cut small pieces off larger pieces.

- *Never* change a blade or a bit unless the power cord is unplugged. Do not depend on the switch being off; you might accidentally hit it.

- *Never* work in insufficient lighting.

- *Never* work while wearing loose clothing, hanging hair, open cuffs, or jewelry.

- *Never* work with dull tools. Have them sharpened, or learn how to sharpen them yourself.

- *Never* use a power tool on a workpiece that is not firmly supported or clamped.

- *Never* saw a workpiece that spans a large distance between horses without close support on either side of the kerf; the piece can bend, closing the kerf and jamming the blade, causing saw kickback.

- *Never* support a workpiece with your leg or other part of your body when sawing.

- *Never* carry sharp or pointed tools, such as utility knives, awls, or chisels in your pocket. If you want to carry tools, use a special-purpose tool belt with leather pockets and holders.

HOME WIRING

Electricity is the rapid flow of energy transmitted by electrons. A public, or sometimes private, utility generates the electricity and sends it to your home through overhead or underground wires called service conductors. The flow must make a complete circuit from the utility's generating station, along the lines to your home, through your household circuits and back to the utility.

How Electricity Works

The electricity goes through a meter, usually attached to the outside of the house into the main, service entrance panel. The meter measures how much electricity your home uses during a certain period and you are charged accordingly. At the service entrance panel, which contains a fuse or circuit breaker or fuse/breaker system, the electricity is divided into branch circuits. The fuses or breakers protect these individual circuits.

The branch circuits supply safe electrical power to the various rooms in your home: kitchen, bathrooms, living areas, bedrooms and so on. Each circuit is protected by its own fuse or circuit breaker and is independent of the others. That is why, when something causes one circuit to fail with a blown fuse or tripped breaker, the remaining circuits are unaffected and continue to supply power to the other rooms.

The force that moves the energy is called voltage. The flow itself is called current. The direction of flow changes 60 times a second. Thus, we speak of 120- or 240-volt, 60-cycle alternating current (AC).

Two- & Three-Wire Systems

Most homes built before 1941 had two-conductor (two-wire) electric service. If you live in a home built during this time and the electrical service has not been remodeled, your home may have two-conductor service. In effect, one conductor carries 120-volt current and the other provides a return path. Actually, the current flow alternates in direction, along both conductors.

Two-conductor service may limit the number and type of electrical appliances you can use. Even if the utility ran a third line to increase your service, your existing circuits might not let you use many of today's electrical conveniences. However, it may well be possible to add new circuits capable of handling the current demands of new appliances. Consult your power company and a qualified electrician to determine whether your present service can handle an increased demand.

Most homes have three-conductor service. Two of the wires are always "hot;" meaning power is always present. The third wire, often inaccurately called "neutral," is hot only when current is flowing. In a modern home, where appliances are running all day and night, the neutral wire is always hot. There are 120 volts between each hot wire and the "neutral" conductor, and 240 volts between the two hot conductors. Thus, there is power for lights and small appliances that require 120 volts, and for large appliances that require 240 volts.

Wattage Ratings

To calculate the wattage (power) available in a circuit, first determine its amperage (amp rating). It will be marked on the circuit breaker or fuse for that circuit in the service entrance panel—15 or 20 amps for most room circuits, 30 or 50 amps for most heavy-duty circuits. Then, Watts = Volts x Amps. Thus, a 15-amp circuit with 120 volts carries (15 x 120 =) 1,800 watts; a 20-amp circuit carries 2,400 watts.

The wattage of any one appliance (see chart) should not be more than 80 percent of a circuit's total wattage capacity. Appliances with large motors, such as air conditioners or refrigerators, should not exceed 50 percent of circuit capacity. To operate properly and safely, each such appliance must have a circuit to itself.

Typical Wattage Ratings

Appliance	Rating	Appliance	Rating
Air conditioner unit	800-1500	Garbage disposer	500-1000
Central air conditioner	5000	Hair dryer	400
Electric blanket	150-500	Heater (portable)	1000-1500
Blender	200-400	Heating pad	50-75
Broiler (rotisserie)	1400-1500	Hot plate	600-1000
Can opener	150	Hot water heater	2500-5000
Clock	13	Microwave oven	650
Clothes dryer (240-v.)	4000-5000	Radio	10
Clothes iron (hand)	700-1000	Range (per burner)	5000
Coffee maker	600-750	Range oven	4500
Crock pot (2 quart)	100	Refrigerator	150-300
Dehumidifier	500	Roaster	1200-1600
Dishwasher	1100	Sewing machine	60-90
Drill (hand)	200-400	Stereo	250-500
Fan (attic)	400	Sun lamp	200-400
Fan (exhaust)	75	TV (color)	200-4500
Floor polisher	300	Toaster	250-1000
Food freezer	300-600	Toaster-oven	1500
Food mixer	150-250	Trash compactor	500-1000
Fryer (deep fat)	1200-1600	Vacuum cleaner	300-600
Frying pan	1000-1200	Waffle iron	700-1100
Furnace (gas)	800	Washing machine	600-900
Furnace (oil)	600-1200		

The Working Tools

Electrical projects require several specialized tools. They also require several standard hand and power tools, such as hammers, chisels, squares and portable electric drills. You may already have many of these tools. However, if you do not, they may be purchased or rented on a project-to-project basis.

In order to save time, money and frustration, we recommend that you buy quality tools and equipment at the outset. Taken care of properly, they should provide many years of service.

The tool and equipment needs have been organized into two categories: basic tools needed for most repairs and simple projects, and additional tools for more ambitious projects, especially those involving carpentry skills.

You will find a selection of electrical tools and equipment at many home center stores and hardware outlets.

The Basic Tools

- Continuity tester or voltage tester.
- Multipurpose wire stripper. Removes insulation without damaging wire. Some strippers also cut and bend wire.
- Needlenose pliers. Excellent for bending tight loops in wire to go around terminals.
- Lineman's pliers. These 7- and 8-inch pliers have flat jaws used to bend, pull, twist and grip wires. Some have wire and cable cutters.
- Set of standard screwdrivers.
- Set of Phillips screwdrivers.
- Set of nut drivers.
- Pocket knife or utility knife.
- Adjustable wrench. Buy either an 8- or 9-inch wrench for tightening nuts and connectors.
- Cable insulation ripper. Removes cable insulation.
- Electric solder gun.
- Volt-ohmmeter for circuit tests.
- Tape measure.

Additional Tools

Many electrical projects require carpentry tools. The tools frequently needed are listed below.

- Locking pliers. Used to grip and hold wires, tighten bolts and pull cable through conduit and holes.
- Diagonal, cable or side cutters. Used to cut wires in cramped quarters, such as outlet boxes.
- Adjustable pliers. Used to handle locknuts and cable connectors.
- Pipe wrench. A 10-inch wrench is used for working with conduit.
- Variable-speed portable electric drill with several hole-saw, masonry and wood bits.
- Wood chisel set; cold chisel.
- Portable electric saber saw.
- Carpenter's level.
- Hammer.
- Hacksaw.
- Compass (keyhole) saw.
- Fish tape. Used to pull wires through finished walls and ceilings.

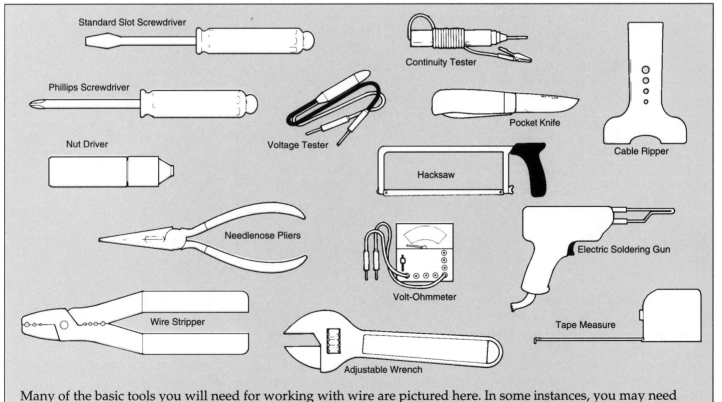

Many of the basic tools you will need for working with wire are pictured here. In some instances, you may need additional tools which are listed above.

Work with a helper. One person works at the main service panel; while the other person tests switches, outlets and appliances.

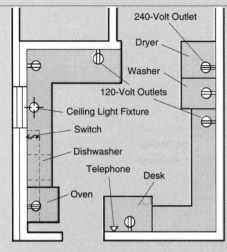

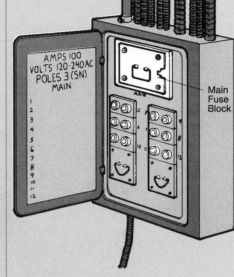

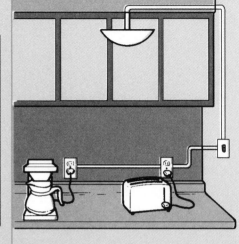

1 *Sketching the Floor Plan.* Draw a floor plan of each room. Mark every receptacle, switch and light fixture. Sketch in heavy equipment that is connected to the service panel. If this equipment has its own separate circuit, so note it.

2 *Numbering the Circuits.* At the main service panel, number each circuit breaker or fuse with a stick-on or glue-backed label.

3 *Setting Up the Test.* Work on one room at a time. Turn on all lights and appliances in that room. Plug in any lamps. If the room has a double receptacle, plug a light into each receptacle. Do not turn on heavy equipment.

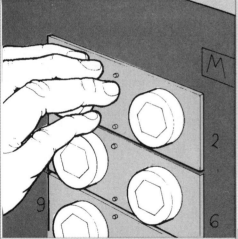

4 *Recording the Circuit.* The person at the main service panel now turns off the first breaker or fuse. On the floor plan, the helper records the number of the circuit or fuse next to each affected switch, light fixture and receptacle.

5 *Identifying Other Circuits.* Turn on the first circuit breaker or fuse and turn off the next one. Continue the recording procedure until every switch, light fixture and receptacle in the room has been labeled with its circuit number. Do this in every room.

6 *Adding Heavy Equipment.* When the circuits controlling all lights, switches and receptacles have been identified, identify and label the circuits serving the heavy appliances.

Mapping Circuits

The starting point for all electrical repair and improvement projects is the main electrical service panel, usually called the fuse box or the circuit breaker panel. This is where all circuits start and end. When there is trouble on a circuit, such as an overload or short, the fuses or breakers shut off the power at this point. When you do any work on a circuit, you must first remove a fuse or trip the breaker to turn off the power at this stopping and starting location.

The following two circuit protectors may not be found in the main service panel:

■ A separate fuse box or circuit breaker added to the main electrical system to power a major circuit or appliance. For example, some older homes have a separate circuit to power a central air-conditioning system. You will find the location of this fuse box when you map the circuits in your home.

■ Some appliances have a built-in protective system with a fuse or circuit breaker. An example is a garbage disposer that has a power-overload device. When an overload occurs, the built-in system shuts down only the appliance. You push a reset button on the appliance to restore power to it. Appliances and devices with similar protection include ranges, clothes dryers, ground fault circuit interrupter (GFCI) outlets and heavy-duty motors.

The built-in system protects only the appliance; the circuit to which it is connected is protected by its own fuse or circuit breaker in the main service panel—one does not substitute for the other. In general, the built-in protector is designed to trip and shut down the appliance at a lower overload level than the fuse or breaker that protects the entire circuit. This helps prevent an appliance problem from affecting the overall circuit wiring.

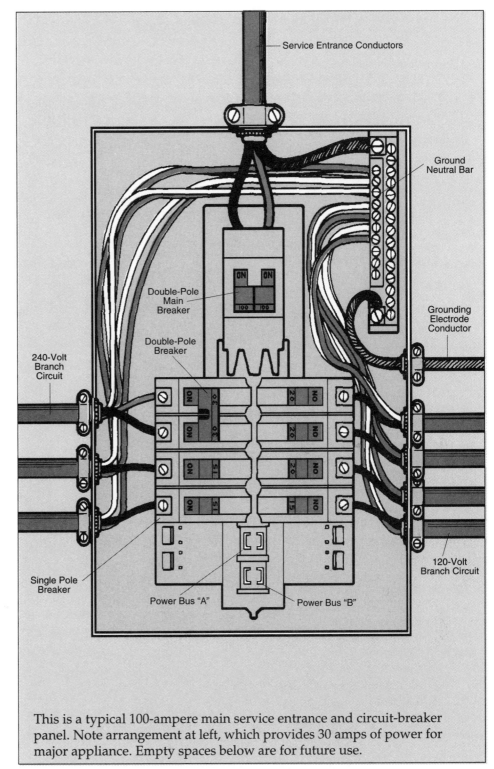

This is a typical 100-ampere main service entrance and circuit-breaker panel. Note arrangement at left, which provides 30 amps of power for major appliance. Empty spaces below are for future use.

Different Designs. Some circuit systems use only fuses, while others use only toggle-type circuit breakers. There also are systems that combine fuses and breakers. Fuses and toggles come in somewhat different shapes and sizes.

Circuit Protection. Fuses and circuit breakers interrupt the current flow in situations where circuit overloading or line-to-line or line-to-ground faults have occurred.

Before you replace a fuse or reset a circuit breaker, find out what caused the power shutoff and correct this condition. Although installing a new fuse or resetting the breaker may restore power to the circuit, it will be temporary; the circuit will shut down again fairly quickly unless you correct the trouble. The problem may be an overloaded circuit, a short circuit in a damaged wire or a broken circuit in an appliance.

Sometimes, but very, very infrequently, a problem may exist in the main service panel. If, after making a thorough check, you can not find trouble on the circuit, suspect the service panel. However, do not try to repair any damage yourself, unless you have the know-how. Hire a licensed electrician to do the job.

Safety at the Panel. When replacing a fuse or resetting the toggle on a circuit breaker, work safely by standing on a completely dry piece of plywood or a short length of 2x6 or 2x8 lumber. Here is a way to be sure the piece of lumber is there when you need it: Cut a piece of wood and drive a nail into the edge or end of it; then bend the nail over into a hook shape and hang the wood on a pipe or similar object near the main service panel.

Wear safety goggles and always use fuse pullers to remove or replace cartridge fuses. Use just one hand to remove or replace plug fuses or reset circuit breakers. This will avoid creating a circuit between you and the panel.

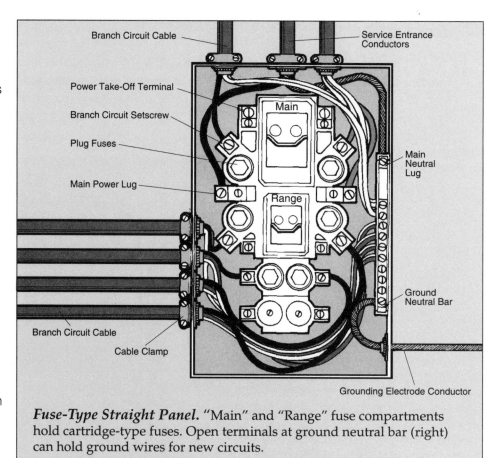

Fuse-Type Straight Panel. "Main" and "Range" fuse compartments hold cartridge-type fuses. Open terminals at ground neutral bar (right) can hold ground wires for new circuits.

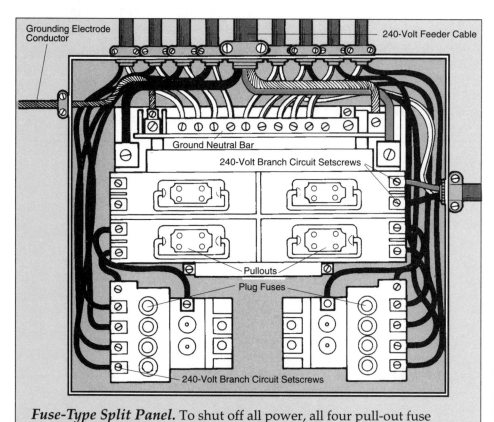

Fuse-Type Split Panel. To shut off all power, all four pull-out fuse compartments (in top section) must be removed. Screw-in plug fuses (below) control the individual circuits.

Overloads & Shorts

Plug Fuses. An overload occurs when too many appliances and lights on a circuit demand more current than the circuit can deliver safely. In this case, the small wire in the fuse will break without heating excessively. The window will be clean, and you should be able to see the broken wire.

A short circuit occurs when a bare wire carrying electricity touches another bare wire carrying electricity or touches the grounded metal case of an appliance. The rate of the current flow quickly becomes excessive. This in turn produces heat, which destroys the fuse wire. The fuse wire vaporizes and sprays the fuse window with discolored material.

Circuit Breakers. Circuit breakers are protective switches that automatically flip off when there is an overload or short circuit. You reset a circuit breaker by pushing the switch or toggle to the full OFF position, and then to the full ON or RESET position.

Since you do not have a window in a circuit breaker to help determine the cause of a short circuit or overload, make a list of the lights and appliances that were operating on the circuit when the breaker tripped. Then add up the total wattage you were pulling at the time of the power failure and divide the wattage by the voltage. If the resulting amperage figure is more than the capacity marked on the failed fuse, there was an overload.

Cartridge Fuses. Two types of cartridge fuses are used in homes. The round-ended type, with a capacity of 10 to 60 amps, is used to protect circuits that supply a major appliance.

The other type of cartridge fuse is usually used in residential installations to protect the main power circuit. This fuse has knife-blade end contacts and is rated at 70 to 600 amps. The two types cannot be interchanged.

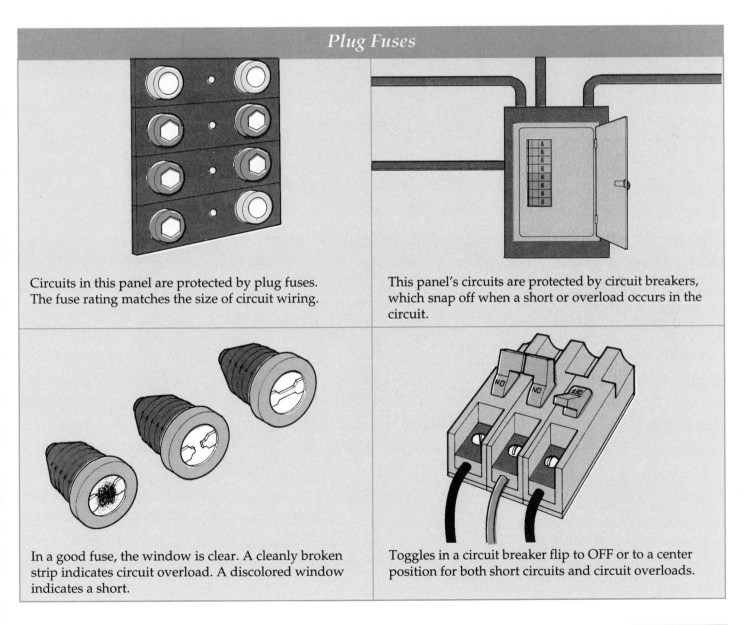

Plug Fuses

Circuits in this panel are protected by plug fuses. The fuse rating matches the size of circuit wiring.

This panel's circuits are protected by circuit breakers, which snap off when a short or overload occurs in the circuit.

In a good fuse, the window is clear. A cleanly broken strip indicates circuit overload. A discolored window indicates a short.

Toggles in a circuit breaker flip to OFF or to a center position for both short circuits and circuit overloads.

Changing Cartridge Fuses

1 **Shutting Off Power.** Some service panels with cartridge fuses are controlled by a lever along the outside edge of the panel. Move the lever to the OFF position. Then open the box.

2 **Removing the Fuse.** Using a fuse puller, grasp the middle of the fuse and pull it out from the spring clips. If the fuse has knife-blade ends, do not bend them.

3 **Compartment Fuse.** Some cartridge fuses are mounted in a compartment-type housing. To remove the fuses, grasp the wire-loop handle and pull the compartment straight out of the panel.

4 **Testing the Fuse.** Touch one probe of a continuity tester to one end of the fuse and the other probe to the other end. If the tester lights, the fuse is okay. If the tester does not light, replace the fuse. To install a new cartridge fuse, push it into the spring clips by hand.

Caution*: Always remove a fuse from the service panel before testing it.*

1 To turn off the power, pull the lever on the outside of the box. Then open the cover of the box.

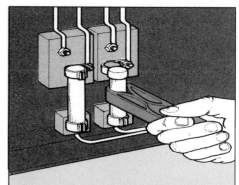

2 With a cartridge fuse puller, remove the fuse from the spring clips that hold it tightly in position.

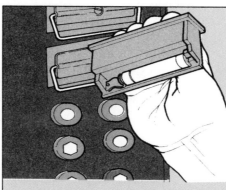

3 Typically found in appliance circuits, fuses are removed after compartment is pulled from panel.

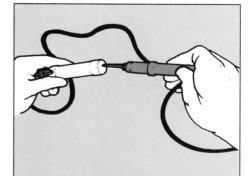

4 Touch continuity tester probes to fuse ends. If test lights, fuse is okay and may be reinserted in box.

Special-Purpose Fuses

Time-Delay Fuse. This fuse is used in circuits that supply heavy appliances, such as air conditioners, that demand a temporary surge of power when turned on. It is made with a spring-loaded metal link attached to a solder plug. Instead of blowing immediately, the plug begins to melt. It must melt through completely before the fuse will blow.

Type S Fuse. This fuse is made up of two parts; an adaptor screws into the socket in the main fuse panel and the fuse screws into the adaptor. The threads of a specific amperage fuse are designed to screw into threads of the same amperage adaptor and no other (you cannot screw a 15-amp Type S fuse into a 20-amp Type S adaptor).

Circuit Breaker Fuse. These fuses have a push button that pops out from the center of the face. When the fuse blows, you simply push in on the push button to reset the fuse. It works like a toggle-type circuit breaker.

WORKING WITH WIRE

Technically, the metal through which electricity flows is called a conductor. In the real world, it is called wire, cord and cable.

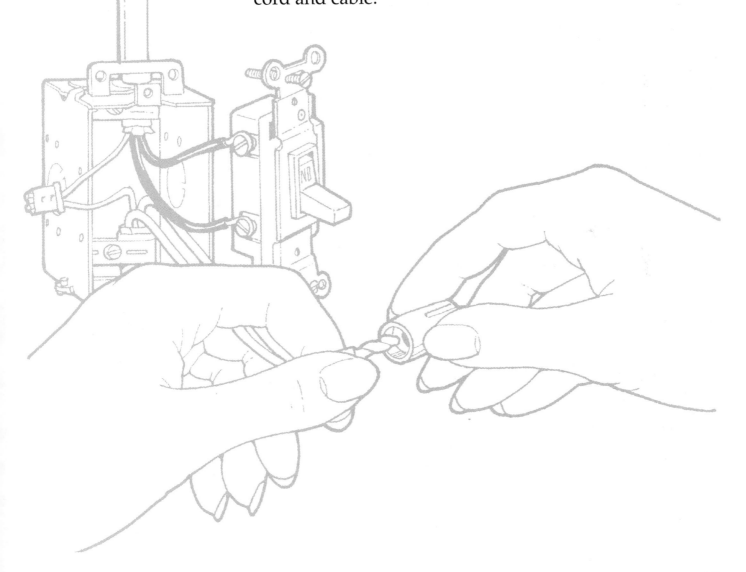

Wire Basics

For practical purposes, a wire is a single strand of conductive material enclosed in protective insulation. Single-strand wire can be purchased off of a roll at any length you need. Sometimes it can be bought precut and packaged in standard lengths. A cable has two or more wires grouped together within a protective sheathing of plastic or metal. Cable is normally sold boxed in precut lengths of 25, 50 or 100 feet. Cord usually is a series of stranded wires encased in insulation. Cord is sometimes precut and packaged, but is usually sold off the roll. All conductors are priced by the lineal foot.

There are three types of wires: copper, copper-clad aluminum and aluminum. For any project, you always should use the same type of wire that is installed in your home. You can determine this by opening a switch or outlet box, pulling out the wires and noting the information printed on the insulation. The markings tell you the voltage, the type of wire or cable, the manufacturer and the AWG wire size.

Aluminum Wire. Aluminum wire requires special care. It does not behave like copper wire. Aluminum wire tends to expand and contract, working itself loose from terminal screws. This can cause trouble—mainly electrical fires. If your home uses copper-clad aluminum wire, do not add aluminum wire to it. Use copper or copper-clad aluminum wire.

If your home has aluminum wire, check to make sure that the switches and receptacles are marked CO/ALR or CU/AL. The CO/ALR marking is used on switches and receptacles rated up to 20 amps. The CU/AL marking is used on switches and receptacles rated at more than 20 amps. If the switches and receptacles do not bear these markings, replace them with those that do.

Never use aluminum wire with any back-wired switch or receptacle that requires pushing the wire into the device. Aluminum wire must connect to terminal screws.

Since recommendations for wire sizes are generally for copper and copper-clad aluminum wires, you must readjust the designation to the next larger size when using aluminum wire. For example: If No. 14 (copper) wire is recommended and you are using aluminum wire, you must use No. 12 wire instead.

Wire Size. You will probably be concerned mostly with No. 14 and No. 12 wire sizes. The term wire refers to a single conductor. In a cable containing two wires, both wires will be the same size.

Wire numbers are based on the American Wire Gauge (AWG) system, which expresses the wire diameter as a whole number. For example, No. 14 AWG wire is 0.064 inches in diameter, and No. 12 AWG is 0.081 inches. The smaller the AWG number, the greater the diameter and the greater the current-carrying capacity. The National Electric Code requires a minimum of No. 14 AWG wire for house wiring. Exceptions to this are the wiring used in lighting fixtures, furnace controls, doorbells and other low-energy circuits.

Wire Ampacity. You also must consider the wire's ampacity, or the current in amperes that a wire can carry continuously under conditions of use without exceeding its temperature rating.

If a wire is too small for the job, it will present a greater-than-normal resistance to the current flowing around it. This generates heat and can destroy insulation, which can cause a fire.

No. 12 wire is rated to carry a maximum of 20 amps; No. 14 wire is rated to carry up to 15 amps.

Cable. House circuits are usually wired with nonmetallic sheathed cable, with metal-armored cable or with insulated wires running through metal or plastic pipe called conduit.

For most projects, you will be working with flexible nonmetallic sheathed cable known by the trade name Romex. It contains insulated power, neutral wires and a ground wire.

Armored cable is called BX. Inside the flexible metal sheathing are insulated power and neutral wires and a ground wire. The use of BX cable sometimes is restricted by code. Check your local codes. BX also is restricted to indoor-use or dry locations. It sometimes is specified for use where power wires need extra sheathing protection.

Conduit, according to code, can be galvanized steel pipe or plastic pipe. There are three types of metal conduit: rigid, which is preferred for outdoor use; intermediate, and electrical metal tubing or EMT, a newer type popular for house wiring. Standard conduit diameters are 1/2, 3/4, 1 and 1¼ inch. There are fittings to join conduit for straight runs and at 45-degree angles. The material is bent with a tool called a hickey.

Cord. This is stranded wires encased in some type of insulation, such as plastic, rubber and cloth.

Zip cord, for example, is two wires, usually No. 18 gauge, encased in a rubber-like insulation and held together with a thin strip between wires. The wires easily come apart by unzipping them; hence the name zip cord. Cord is used for lamps, small appliances, and cord sets that have plugs and/or receptacles on one or both ends of the cord.

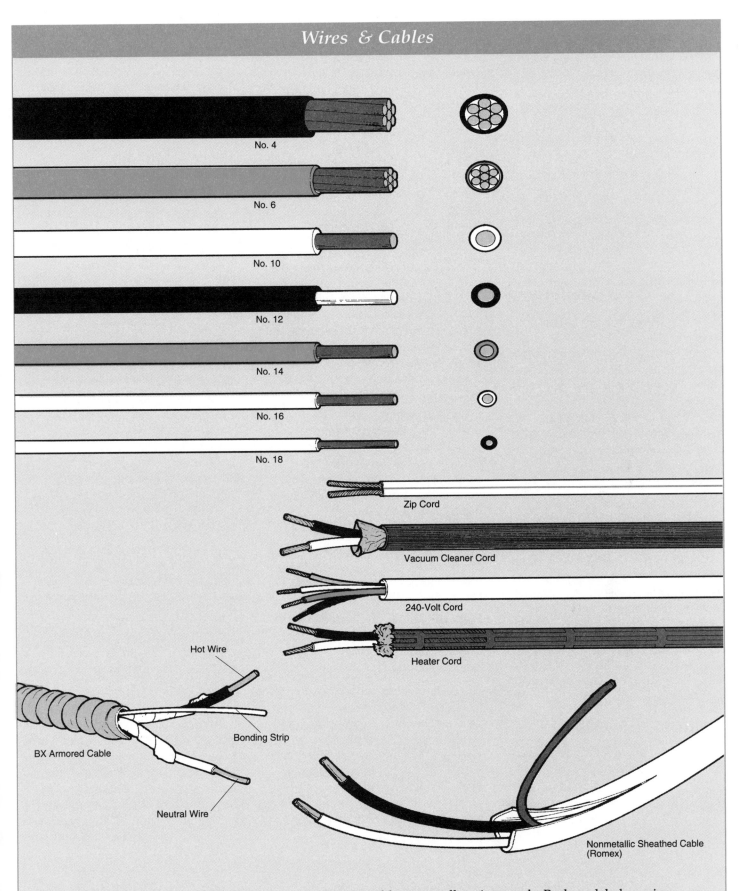

No. 4

No. 6

No. 10

No. 12

No. 14

No. 16

No. 18

Zip Cord

Vacuum Cleaner Cord

240-Volt Cord

Heater Cord

Hot Wire

Bonding Strip

BX Armored Cable

Neutral Wire

Nonmetallic Sheathed Cable (Romex)

Meeting Your Needs. Many kinds of wire and cable are sold to meet all project needs. Package label or wire insulation gives data on wire size or gauge, type of insulation and power capacity.

Selecting Wire by Insulation

Wire, both solid, single strands and cable, is available with many different types of insulation. You must select the right wire for the location in which it will be used.

Wire Specifications

The most commonly used types are described below. They are: RHW; T,TF, and TW; THHN; THW and THWN; and XHHW.

RHW. The insulation is moisture- and heat-resistant rubber. The wire may be used in either wet or dry locations.

T,TF and TW. The T stands for thermoplastic. For household electrical projects, you probably will use more of this type of wire than any other. However, it is only used for dry locations. Type TF is a moisture-resistant thermoplastic insulation that may be used in place of TW in both moist and dry locations.

THHN. A flame-retardant, heat-resistant insulation specified for dry and damp locations. Because it is thin, THHN often is used in conduit to allow more wires to be installed.

THW and THWN. A flame-retardant, moisture- and heat-resistant thermoplastic insulation for use in wet or dry locations.

XHHW. This insulation is a flame-retardant cross-linked, synthetic polymer. It is specified for use in dry and damp locations as well as wet locations.

Cable Specifications

The types of sheathed cable that are most commonly used are: NM, NMC and UF.

Type NM Cable. This cable is for use only in dry locations. It is used most often in house circuits. Each wire (with the possible exception of the equipment grounding conductor) is wrapped in its own plastic insulating sheath. The three wires are then wrapped in a paper insulator, and then covered with plastic.

The wire in Type NM cable is either AWG No. 12 or AWG No. 14 for normal house circuits. Larger sizes, such as No. 10 or more, are used for heavy appliances. The National Electrical Code specifies that No. 12 wire must be used for certain household circuits.

In either size, NM cable is available with two or three conductors plus an equipment-grounding conductor. This ground-wire system is highly recommended. Use three-conductor NM cable for heavy-duty circuits, especially where two hot wires are needed for the hookup.

Type NMC Cable. This cable may be used in both damp and dry locations. The distinguishing characteristic of this cable is that the individually insulated wires are embedded in solid plastic to provide protection against moisture. As a result, it is appropriate for basement installations where codes permit.

Type NMC is available with two or three conductors plus an equipment grounding conductor and in AWG No. 12 and AWG No. 14.

Type UF Cable. This cable is used in wet locations, including burial underground. UF cable may be used instead of conduit.

The distinguishing characteristic of this cable is that the individually insulated wires are embedded in water-resistant solid plastic that is heavier than that used in Type NMC cable. The UF cable is available in AWG No. 12 and AWG No. 14 as well as other size wires. It contains two or three conductors plus an equipment grounding conductor.

Abbreviations on Wire

Markings on the insulation, plastic sheathing and on nonmetallic cable explain what is inside and identify the type of insulation covering. For example consider the following designation:

14/2 WITH GROUND, TYPE NMC, 600V (UL).

The first number tells the size of the wires inside the insulation or cable; in this case No. 14 gauge. The second number tells you that there are two conductors (wires) in the cable. There also is an equipment grounding wire, as indicated. The type of cable is given; followed by the number that indicates the maximum voltage allowed through the cable.

Finally, the UL (Underwriters' Laboratories) notation assures you that the cable has been rated as safe for the uses for which it was designed. The National Electrical Code requires that wires of types NM and NMC have a rating of 90° Centigrade (194° Fahrenheit).

Estimating Wire Needs

To estimate the amount of wire or cable you will need for a project, measure the distance between the new outlet and the power source.

Add an extra foot for every connection you will make. Then, to provide a margin for error, add 20 percent to this figure.

For example, if you measure 12 feet between a new and existing receptacle, add another 2 feet for the two connections, making a total of 14 feet. Then add 20 percent (about 3 feet) to the total. To do the job, start working with 17 feet of cable. The same formula is used for wire, with the exception of lamp/appliance cord.

Wire (Cable) Connectors

Cable, or more specifically the wires inside the cable sheathing, may be connected in several ways: with wire connectors, crimped connectors, split-bolt connectors, and solder.

The code specifies that all connections (splices) must be made inside a box.

Wire Connectors

Wire connectors or connectors that do not use solder are sized according to the size of the wires to be spliced. They are used widely and are code-approved.

Crimped Connectors

These fasteners are similar to wire connectors, but they sometimes are not permitted by code in local areas. Check your local code.

To use crimped connectors, strip about 1/2- to 3/4-inch of insulation off the wires to be connected. Twist the wires together with pliers so the joint is well wrapped and tight. Then insert the wire ends in the connector and crimp the end of the connector with a tool made especially for this job.

Split-Bolt Connectors

These connectors are available in assorted sizes to correspond with wire sizes. Split-bolt connectors are basically designed to be used with the larger wire sizes—from No. 6 gauge and larger.

Strip about 1/2- to 3/4-inch insulation off the wires to be joined. Thread the wires into the connector loop and tighten the nut with pliers or an adjustable wrench or nut driver. Then wrap the splice with several layers of plastic electrical tape.

Solder Connections

You may solder wires together for a strong, tight splice. It is time-consuming to do so, however.

Use only rosin-core solder to create the soldered splice. Wrap the splice with electrical tape to match wire insulation.

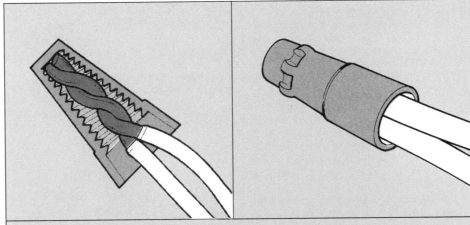

Wire Connectors and Crimped Connectors. The plastic wire connector (left) is screwed onto the twisted wire connection. Crimped connectors (right) work like wire connectors, except that the ends are crimped with a special tool after the wires are inserted.

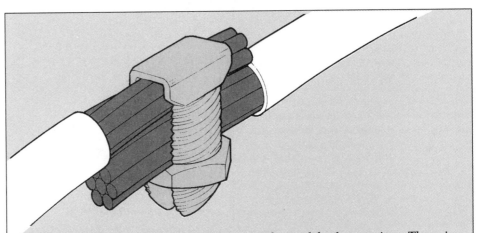

Split-Bolt Connectors. This type is mainly used for large wires. The wire ends are stripped and then slipped into the connector. A nut on the connector compresses the wires, making a very strong splice. Wrap the splice with plastic electrical tape.

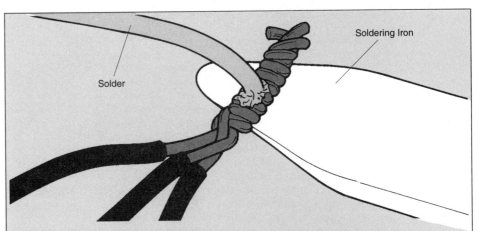

Soldering Iron

Solder

Solder Connections. For soldered connections, use rosin-core (noncorrosive) solder. Lay the wires on the soldering gun (iron), let the wires heat and then apply the solder to the splice. Wrap the splice with plastic electrical tape to the thickness of the wire insulation.

How to Strip Off Wire Insulation

To remove the insulation from wires you can use a jackknife, but an inexpensive wire stripper is a better tool. First cut the wire to the right length. About 3/4 inch of insulation should be stripped off the wire for the best terminal connection.

1 **Matching the Wire to the Stripper.** Put the wire in the hole in the handle that matches the wire size. For example, for a lamp wire, the hole will be No. 18 or No. 16.

2 **Rotating the Stripper.** Lightly grip the handles of the stripper in a closed position with the wire inserted in the correct hole. Then rotate the stripper around the wire a couple of times.

3 **Pulling Off the Insulation.** With the handles still closed, pull the wire out of the stripper. The handles will grip the insulation and the pulling action will strip it off.

4 **Tapering the Insulation.** There are special wire strippers that provide a tapered cut. This is preferred over a square cut.

Using a Knife

If you use a jackknife to remove the insulation, be very careful to cut only the insulation and not the wire. Cut completely around the insulation and then pull it off with your fingers.

Stripping Cable Insulation

You can buy a stripping tool to slice the insulation on cable. Once stripped, the insulation then has to be trimmed with a knife or scissors.

You also can use a utility knife to make the first cut. Be extremely careful that you do not cut the insulation on the wires inside the cable as you slice the outside insulation covering.

For most connections, you will need to strip back the outer insulation about 3 to 4 inches.

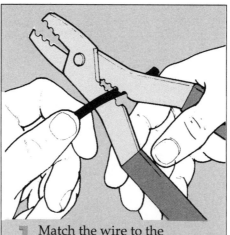

1 Match the wire to the numbered hole in the handle of the wire stripper. Insert the wire in this hole; grip handles.

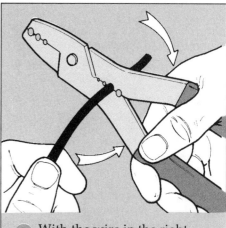

2 With the wire in the right hole, lightly grip the stripper and rotate the stripper completely around the wire.

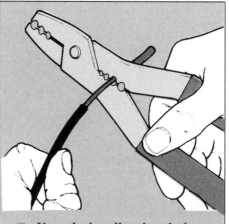

3 Keep the handles closed after the insulation is cut through. Then pull the wire out of the tool to strip off insulation.

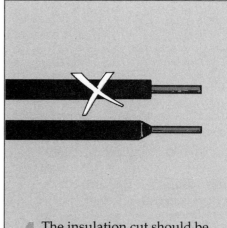

4 The insulation cut should be tapered, not square, if possible. Some wire strippers provide a tapered cut.

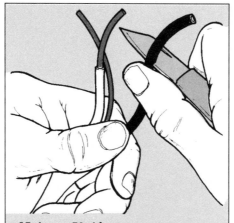

Using a Knife. Be careful not to cut or nick the wire. A glove or thumb protector is a wise precaution against cuts.

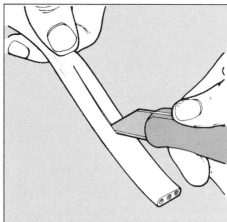

Stripping Cable Insulation. A cable stripper or knife removes outer insulation from cable. Do not cut wire insulation.

Making Wire Splices

According to the code, all wire splices must be enclosed in a switch, outlet, fixture, or junction box.

Stranded Wires

Strip off about 3/4 inch of insulation. With your fingers, twist each wire individually so the strands are tightly together. Then, with your fingers, twist the two wires together.

Solid to Stranded Wires

Strip off about 3/4 inch of insulation from both wires. Twist the stranded wire tightly, then wrap it around the solid wire with your fingers. Then with pliers, bend over the solid wire to secure the stranded wire to the solid wire.

Solid to Solid Wire

Strip off about 3/4 inch of insulation from both wires. With pliers, spiral one piece of solid wire around the other piece, making the twist fairly tight, but not tight enough to break the wire.

Stranded Wires. Twist stripped, stranded wire with fingers, make a tight wire. Then twist the wires together tightly.

Solid to Stranded Wire. Wrap stranded wire around solid wire, using the solid wire as a base. Then bend over solid wire to lock the splice.

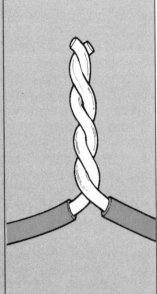

Solid to Solid Wire. Twist solid wires together tightly with pliers. Do not over-tighten the spiraled wires or you will crack or break them.

Twist on Wire Connector

1. Pick a wire connector that fits the splice. Insert the wires into the wire connector with a slight twisting motion. Do not apply too much downward pressure on stranded wire splices. If you do, the wire will buckle and flatten. Just screw the nut onto the splice.

2. After the wire connector is in place, it should completely cover the splice with a bit of wire insulation seated in the opening of the wire connector.

3. When you are satisfied that the splice is tight and securely covered in the wire connector, wrap the wire connector and an inch or so of the projecting wires with plastic electrician's tape. The tape is a safety measure that helps to make a stronger splice. However, do not rely on the tape to hold the slice together. If the splice is not tight and covered by the wire connector, remove the splice and start again.

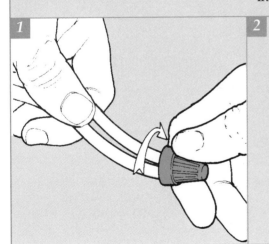

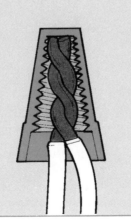

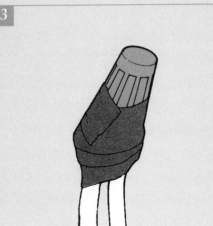

Working with Flexible Clad Cable (BX)

Armor-clad cable, commonly called BX, a trade name, has an outer wrapper of galvanized steel; the cable contains two or three wires. Each wire is individually wrapped in insulation and then with paper. The cable must contain an internal bonding strip that is in intimate contact with the armor for its entire length. Check local codes; the use of BX cable may be prohibited. BX is specified for use in dry locations only.

1 **Cutting the Armor.** Use a hacksaw to cut the flexible steel wrapper about 8 inches from the end.

Make the cut diagonally across one of the metal ribs, and stop sawing just as soon as the blade cuts through the metal. Do not cut any deeper or you will damage the wires inside the wrapper.

Then, with your fingers, bend the cable back and forth until the metal snaps apart. Slide the armor off the cable.

2 **Removing the Insulation.** Unwind the paper insulation around the wires inside the cable and cut the paper with a knife or scissors. Then remove the plastic insulation. The amount of insulation to be removed depends on what you plan to do with the ends of the wires. For terminals, strip off about 1/2 to 3/4 inch. For splicing, strip 3/4 inch. For push-in terminals, use the stripping gauge on the back of the switch or receptacle.

3 **Inserting the Bushing.** Slip a plastic bushing made for this cable around the wires and into the cut cable opening. The bonding strip goes outside the bushing, between the bushing and cable.

4 **Connecting the Strip.** The bonding strip is sometimes fastened to the cable connector screw.

1 Saw diagonally across the cable using a hacksaw. When the blade has almost cut through steel, bend cable over to break it.

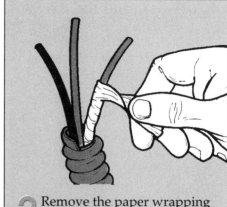

2 Remove the paper wrapping to expose wires inside cable. Then strip wire insulation for connections you'll make.

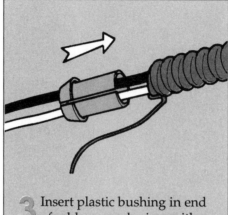

3 Insert plastic bushing in end of cable around wires with bonding strip positioned between bushing and armored cable.

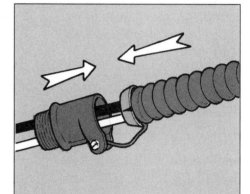

4 Bonding strip is connected to screw tightener on cable connector. Strip is a grounding device for the cable circuit.

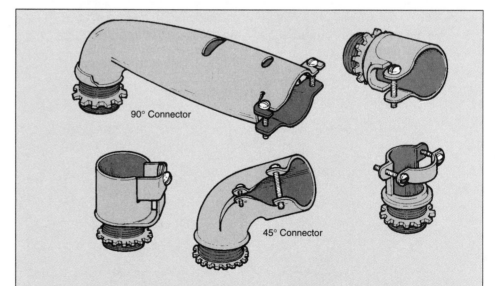

90° Connector

45° Connector

Types of Flexible Armored Cable Connectors. Different types of armored cable connectors are shown here. Some connectors have tiny view slots so electrical inspectors can see that bushings have been installed without disassembling entire connection. You can buy 90- and 45-degree connectors.

Working with Nonmetallic Cable

You probably will buy and work with nonmetallic plastic-sheathed cable more than any other conductor or wire. It is often called by its trade name, Romex. Local codes may allow nonmetallic cable only in certain locations, or may specify that you use another type, such as metallic armored cable, or wires running in conduit.

The outer sheath of nonmetallic cable is usually a moisture-resistant, flame-retardant material. Inside, there are two or three insulated power wires and perhaps a grounding wire.

For most residential wiring, two types of nonmetallic cable commonly are used. They will be labeled Type NM or Type NMC on the package or the cable.

Type NM may be used in dry locations and may be either concealed or exposed. Type NMC meets the same requirements as NM, but in addition it is fungus- and corrosion-resistant. It may be used in moist, damp and corrosive locations such as in hollows of concrete blocks used in building.

Both types contain either copper wire in gauges 14 through 2, or aluminum or copper-clad aluminum in gauges 12 through 2.

Two-Wire/Bare Ground

The cable has a hot wire in black insulation, a neutral wire in white insulation, and a grounding wire, which is bare (no insulation).

Two-Wire/No Ground

This cable consists of two wires: a hot or power wire covered with black insulation and a neutral wire covered with white insulation.

Three Wire/No Ground

This old-style cable has one hot wire covered with black insulation and another hot wire covered with red insulation; the neutral wire has white insulation. A three-wire cable is used to connect three-way switches.

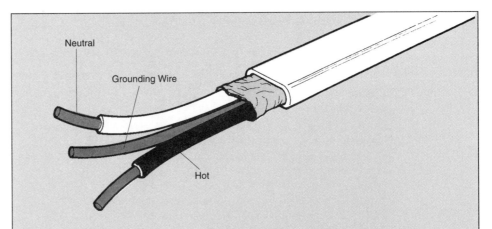

Two-Wire/Bare Ground. Two-wire nonmetallic cable with bare grounding wire has a hot or power wire encased in black insulation. The so-called neutral wire has white insulation; the grounding wire is not insulated. This cable is commonly specified for residential wiring.

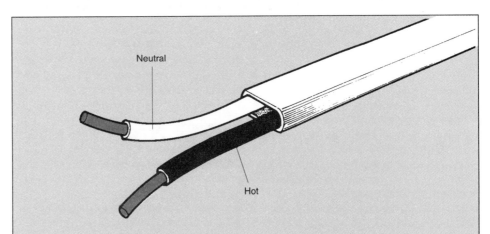

Two-Wire/No Ground. Two-wire nonmetallic cable with no grounding wire has a hot wire in black insulation and a neutral wire in white insulation. Wires in cable are color-coded so they are not mixed when hooking up fixtures along a circuits entire run. Color matches terminals.

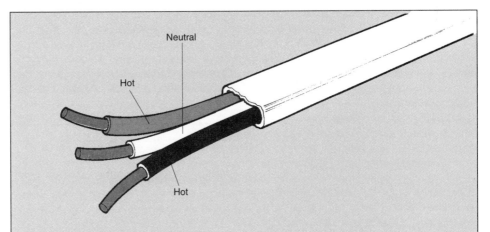

Three-Wire/No Ground. Three-wire nonmetallic cable without a grounding wire has a black-insulated hot wire, a white-insulated neutral wire, and a red-insulated wire that is considered a hot wire. In three-way switch hookups, the red wire becomes the hot switch wire.

Two-Wire/Coded Ground

In this cable, the grounding wire is insulated and often color-coded green or green and yellow stripes. The other color codes are the same: black for hot wires, white for neutral.

Three-Wire/Ground

This cable is commonly used for house circuits in which a grounding wire runs through the complete circuit. The grounding wire may be hooked to a terminal in an outlet box, or it may be connected by a pigtail—a short length of wire—to a grounding terminal in the box or on a receptacle.

Type UF Cable

This type of cable can be used for interior wiring in wet or corrosive locations where type NM cannot be used. It looks the same as other types of nonmetallic cables but is marked with the letters UF on the package or insulation.

Like type UF, type USE cable may be buried underground. It is often used for underground service entrances to buildings.

Cable Wire Sizes

It is recommended that all new residential circuits use No. 12 gauge wire. No. 14 gauge wire may be added to an existing circuit of No. 14 wire.

You will find cable packages and the cable itself marked with the wire size, followed by the number of wires inside the cable sheath. Check the markings carefully so you buy exactly what you need. For example, a cable with two No. 12 wires will be marked "12/2." If there is also a grounding wire, it will be marked "12/2 with Ground." The first cable has two insulated wires, black and another color. The second has those wires and the ground wire either bare, or insulated in green or green and yellow.

Wire gauge Nos. 8-14 designate single wires. Nos. 6-2 are multiple wires held together by the insulation. Nos. 16 and 18 contain multiple strands twisted or braided together.

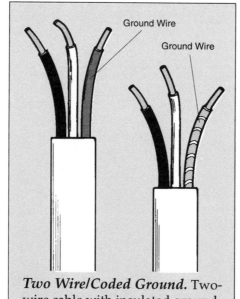

Two Wire/Coded Ground. Two-wire cable with insulated ground wires may be color-coded with solid green or green and yellow stripes.

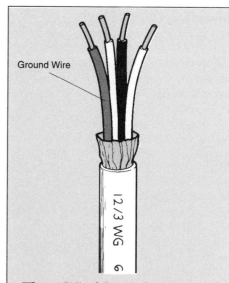

Three-Wire/Ground. Cable for three-way switches has a black-insulated power wire, white neutral wire, red switch wire, and a bare or green ground wire.

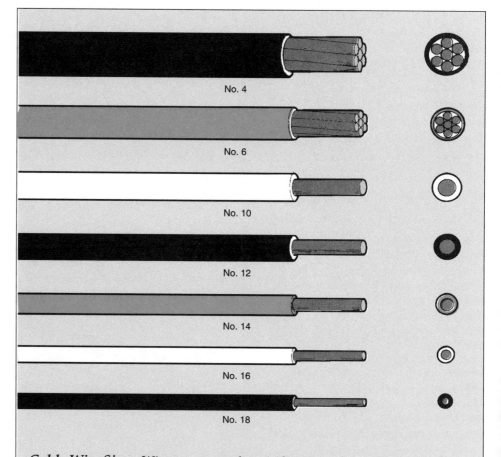

Cable Wire Sizes. Wire gauge numbers indicate conductor sizes; the smaller the number, the larger the wire. Nos. 4 through 8 have multiple wires and use special connectors. No. 16 and 18 wires are used only on fixtures or as extension cords, never for installed circuit wiring.

Installing Nonmetallic or Plastic-Sheathed Cable

To prepare nonmetallic or plastic-sheathed cable for installation, you need a sharp utility knife, a wire stripper and a cable ripper. You can cut the cable with a knife, but a ripper is better in many cases because it protects the insulated wires within the cable sheathing.

1 Cutting the Sheath Insulation. Place the cable on a flat surface, such as a workbench. Measure about 8 inches from the end of the cable and make a mark.

Then insert the cable in a cable ripper at the marked point. Press the cable ripper together with your fingers and pull the cable through the ripper to the end of the cable. If you use a knife instead of a ripper, start cutting the sheath at the mark. Run the knife down the sheath, being extremely careful not to cut the insulated wires inside the cable. It may take several shallow cuts with the knife to part the plastic sheath. If you damage the wire, cut that part off and start again.

2 Removing the Sheath Insulation. With your fingers, peel back the sheath and then use a knife to trim away the excess sheath material at the first cutting mark.

3 Removing the Individual Wire Insulation. With wire strippers, remove about 1/2 to 3/4 inch of insulation from the black insulated power wire, the white-insulated neutral wire, and the green or green and yellow grounding wire (if it is insulated).

Check to make sure that you did not cut the wire insulation with the ripper or knife as you removed the sheath insulation. If you did cut the insulation on the wires, trim off all wires at the cutting mark and start over. The insulation on the wires inside the cable must be completely sound to prevent hazards, such as a short circuit or an electrical fire.

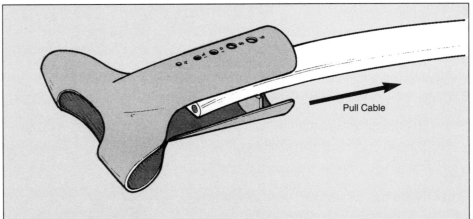

1 Cable ripper makes a clean cut in the outer sheath of nonmetallic electrical cable. Slip the wire into the ripper at the cut-off mark—about 8 in. from the end—and grip the ripper firmly. Then pull the cable through the ripper to make the proper cut.

Pull Cable

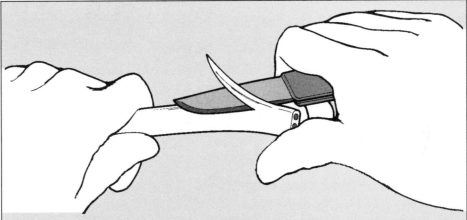

2 Remove excess sheath insulation with a sharp utility knife or jackknife. Trim it flush with the cutoff mark that you made on the cable. Do not nick the inside wire insulation. With a knife, it is wise to wear a glove or thumb protector to avoid cuts.

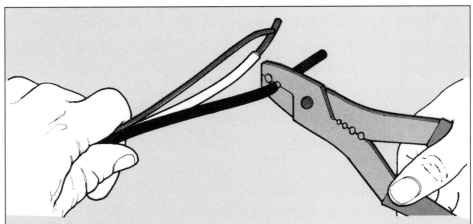

3 Use wire strippers to remove the insulation from the wires inside the cable. If you have accidently cut this insulation with the ripper or knife, trim the wire at the cutoff mark and start the process over again. Damaged insulation can cause electrical hazards.

Working with Aluminum Wire

Most wire that you buy will be copper or copper-clad aluminum wire. You may discover, however, that your home has been wired with solid aluminum and, unless you decide to completely rewire with copper, you will have to work with aluminum wire.

Aluminum wire used to cause problems when it was used in switches, outlets and fixtures that had not been designed for the characteristics of aluminum wire. The wire tends to come loose by expanding and contracting at terminals. Loose wires can cause electric arcing, which can produce electrical fires.

The industry has solved part of the problem with switches, outlets, fixtures and equipment made especially for use with either aluminum or copper wire. The products are plainly marked with the letters CO/ALR.

Wiring Procedures

As described here, extra effort must be made when connecting aluminum wire to terminals. It is also a smart idea to apply the same rules to copper wire when connecting it.

1   **Strip, Loop and Hook.** Remove about 3/4 inch of insulation from the wire. Use wire strippers, if possible. Loop the end of the bare wire with needlenose pliers. Just grip the wire in the jaws of the pliers and wrap it around the jaws, which are rounded. This automatically forms the loop of the size that is required for terminals. Then place the loop around the terminal screw with its opening to the right.

2   **Tightening the Terminal.** When the loop is in place, tighten the terminal screw so the screw and contact plate make full contact with the wire.

3   **Giving a Half Turn.** When the wire is snug under the terminal screw, give the terminal screw another half turn.

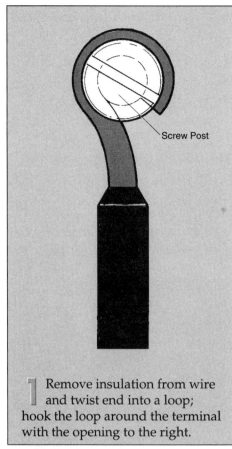

1   Remove insulation from wire and twist end into a loop; hook the loop around the terminal with the opening to the right.

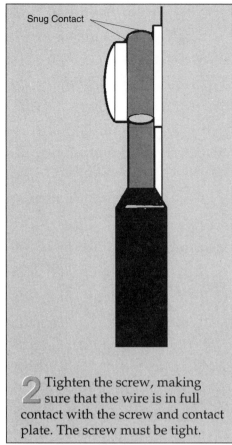

2   Tighten the screw, making sure that the wire is in full contact with the screw and contact plate. The screw must be tight.

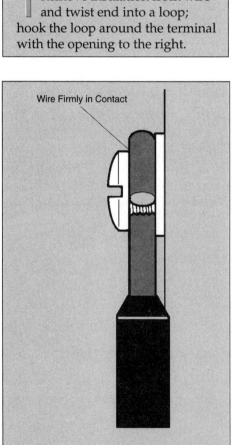

3   Give the screw another half turn after you have initially tightened it. But don't strip slot or threads with too much force.

Copper-Clad Aluminum

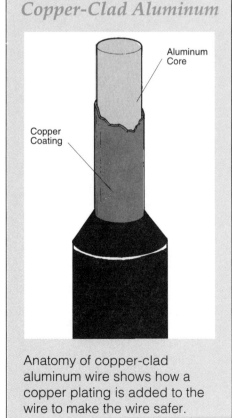

Anatomy of copper-clad aluminum wire shows how a copper plating is added to the wire to make the wire safer.

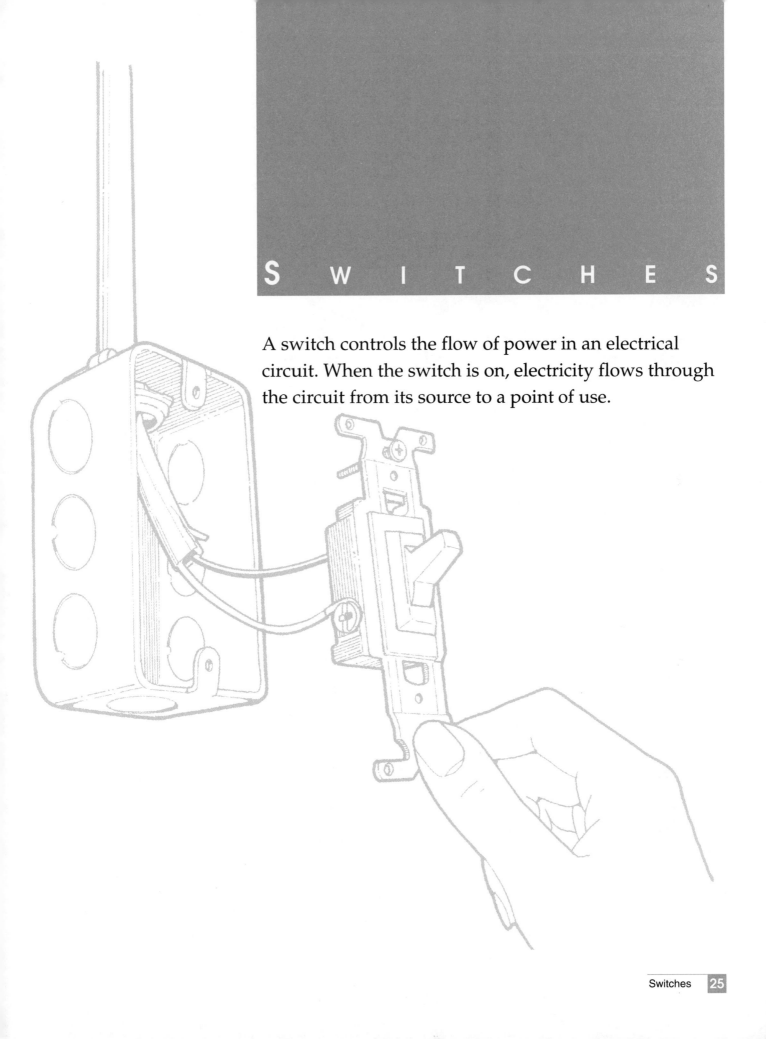

S W I T C H E S

A switch controls the flow of power in an electrical circuit. When the switch is on, electricity flows through the circuit from its source to a point of use.

Types of Switches

Most residential switches are toggle types, also called snap switches. Others include dimmer, pilot-light, time-clock and silent switches.

Single-Pole Switches

A switch with two terminals is called a single-pole switch; it alone controls the circuit. The incoming hot wire is hooked to one terminal screw, and the outgoing hot wire is connected to the other screw.

Three-Way Switches

A switch with three terminal screws is called a three-way switch. One terminal is marked COM, or "common;" the hot wire is connected to this terminal. The other terminals are switch leads. Two three-way switches are used to control a circuit from two places.

Double-Pole Switches

A double-pole switch has four terminals. It is normally used to control 240-volt appliances. A four-way switch also has four terminals. Three four-way switches are used in a circuit to control one outlet or fixture from three separate places. Both switches look the same, but only a double-pole switch has ON-OFF markings.

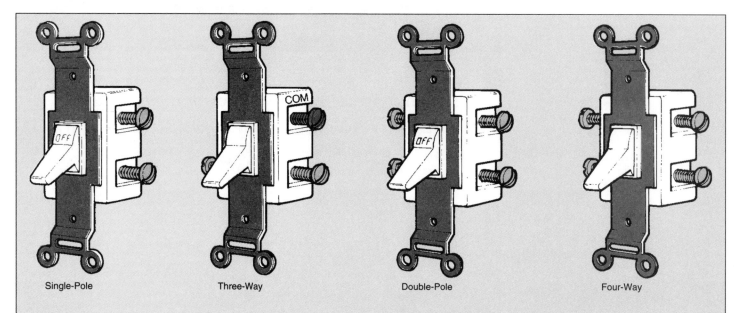

Single-Pole   Three-Way   Double-Pole   Four-Way

Choosing a Switch. Common types of switches include this basic selection for residential housing. Note the difference between four-way and double-pole, which has OFF/ON stamped on the toggle.

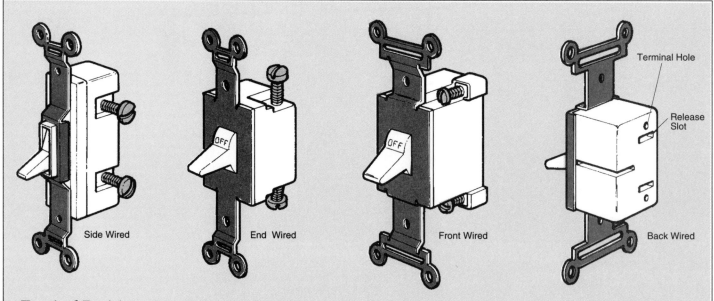

Side Wired   End Wired   Front Wired   Back Wired

Terminal Hole

Release Slot

Terminal Positions. Switches come with different terminal positions for wiring convenience. All those shown here are single-pole. Back-wired switches have terminal holes instead of screws.

Reading the Switch

Switches are stamped with code letters and numbers. Learn how to read these codes so you buy the right products.

UL or UND LAB means that the switch is listed by Underwriters' Laboratories, a testing organization. AC ONLY means that the switch will handle only alternating current. CO/ALR is a wire code indicating that the switch will handle copper, copper-clad, and aluminum wire. 15A-120V means that the switch will handle 15 amperes and 120 volts of power. A new switch must have the same amp and volt rating as the switch it replaces.

Removing Switch Plates

If you have trouble removing an old switch or outlet cover plate, try cutting around the edge of the cover with the tip of a utility knife.

Testing Switches

When you flip a switch and the circuit does not work, the fault may not be in the switch. It could be in the fixture or a fuse. After checking the fuse (see page 11), test the switch, using this procedure:

1 **Shutting Off Power.** Remove the fuse or set the breaker in the switch circuit to OFF.

2 **Testing Terminals.** Remove the switch plate. Touch voltage tester probes to black and white wire terminals to check that the power is off.

3 **Turning the Switch ON.** Touch the continuity tester probe and clip to the terminals. The tester should light with a good switch.

4 **Turning the Switch OFF.** Turn the switch OFF. Touch terminals with a probe and clip. Continuity tester should not light with a good switch.

5 **Testing Mounting Strap.** Fasten tester clip to switch mounting strap. Touch probe to one terminal, then the other and flip switch ON and OFF each time. The tester should not light in any position. If the switch fails, replace it (see following pages).

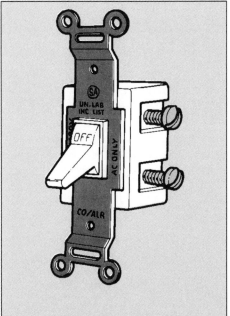

Reading the Switch. Codes are stamped on switches so you know which one to buy for a project. How to read code is explained left.

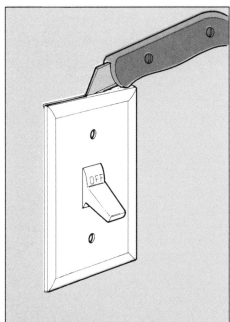

Removing Switch Plates. Paint-sealed plates are easy to remove. Slice the paint seal with a utility knife; then remove the screws.

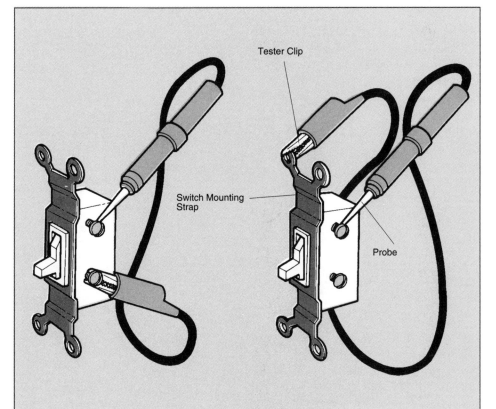

Tester Clip

Switch Mounting Strap

Probe

Testing Switches. There are two basic techniques for testing switches. With the circuit power turned off, connect continuity tester as shown at left. Tester should light with switch ON; should not light with switch OFF. When the probe is connected to switch mounting strap (at either terminal), shown right, tester should not light with switch either ON or OFF.

Working with Single-Pole Switches

It is not difficult to replace or add a single-pole switch. The process may vary slightly, depending on your house wiring and whether the switch is grounded or not. A grounded switch has an extra terminal screw at the base that is green or shows the letters GR. This redundant grounding system is more reliable than systems that do not connect the ground to the switch. If wires are encased in metal conduit, the conduit is usually grounded, but not always.

When replacing a switch or adding a new one, buy switches with a ground-terminal screw, even though it may be necessary to modify your wiring, as explained in this section. Detach only those wires that are connected to the switch itself.

Single-Pole to Fixture

One of the easiest switch/light wiring hookups is a single-pole switch controlling a light fixture. Follow these procedures:

1 Running Wire. Turn off the power at the main service entrance. If the circuit is a new one, run the wire from the service panel to the switch and light, but do not connect it to the panel. Have an electrician do this.

2 Cutting Wire. Cut the wire at the switch.

3 Stripping Wire. Use wire strippers to strip off about 1/2 to 3/4 inch of wire insulation on each end.

4 Connecting Wires. Connect the black wires to the terminals, hooking the wire loops around the terminals in the direction the screw tightens. The white wire bypasses the switch. Connect the grounding wires in the cable from the light and in the power cable to a pigtail (a short wire of the same gauge) that is attached to the grounding terminal in the box. Use a wire connector.

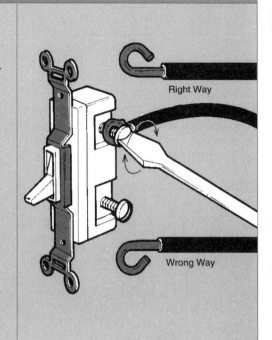

Connect wires to terminals by looping the end around the terminal screw in the direction the screw tightens. This is usually clockwise.

The best way to form a loop in the wire for terminal screws is with needlenose pliers. Strip about 1/2 to 3/4 inch of insulation off the wire end of the wire and bend the bare wire around the jaws of the pliers, forming a perfect loop. Then hook the loop onto the terminals in the direction the screws turn down, and tighten the screws. As the terminals are tightened, the wire is forced under the screw heads and clamped.

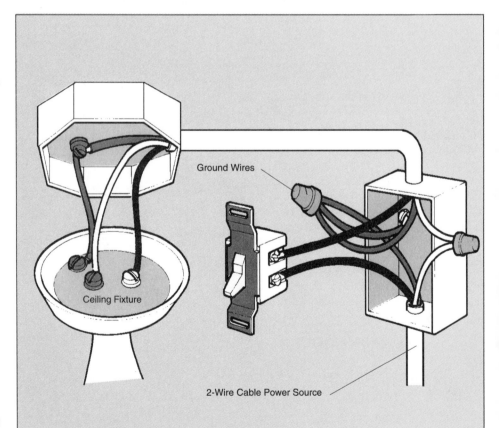

Wiring a Single-Pole Switch to a Light Fixture. A single-pole switch, which controls one light fixture with power coming from the switch, is wired as above. The white wire bypasses the switch; a ground connects to both metal boxes.

Single-Pole Switch Controls Fixture;
Constant Power to Outlet

In this single-pole connection the power is supplied by a two-wire cable with ground. A three-wire cable with ground goes to the light and a two-wire with cable ground to the outlet.

1 **Wiring the Switch.** With a length of black-insulated wire pigtail (same gauge wire), connect the black power wire to the switch and then to the black wire in the three-wire with ground cable. Wrap the wire nut with electrician's tape. Now connect the white wire from the power source to the white wire in the three-wire cable. Use a wire nut and tape it. Connect the red wire in the three-wire cable to the open switch terminal. Finally, connect the cable grounding wires (green or bare) to a pigtail that is attached to the box ground terminal.

2 **Wiring the Outlet.** Connect the black wire to the brass terminal and the white wire to the silver terminal of the outlet, using the two-wire with ground cable. With a pigtail, splice the ground and connect it to the box and the outlet ground terminal.

3 **Wiring the Ceiling Box.** Connect the black wire from the switch to the black wire from the outlet. Add a wire nut and wrap with tape. Connect the grounding from the switch box cable to pigtails from the receptacle ground terminal and the box ground terminal.

4 **Wiring the Light Fixture.** Connect the red wire from the switch to the black light wire, if the light is prewired. If it is not, then connect the red wire to the brass-colored light terminal. The white wire is spliced to the light's white wire or connected to the light-colored terminal on the light.

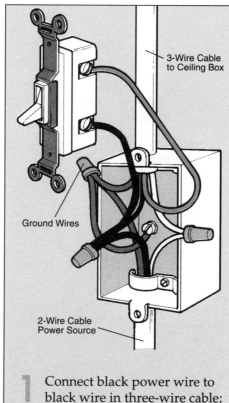

1 Connect black power wire to black wire in three-wire cable; white wire to white wire, red wire to the brass switch terminal.

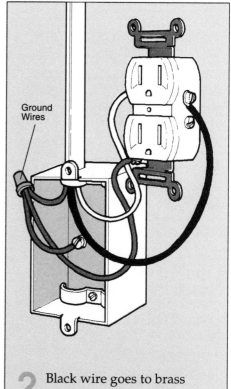

2 Black wire goes to brass terminal on outlet; white wire goes to silver terminal. Ground goes to terminal and box.

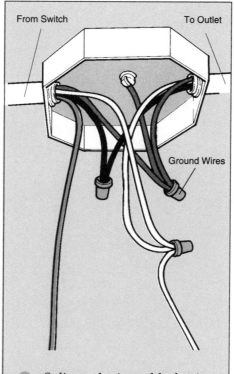

3 Splice red wire to black wire of lamp or to brass terminal. Black wire goes through box to outlet; white to light/outlet.

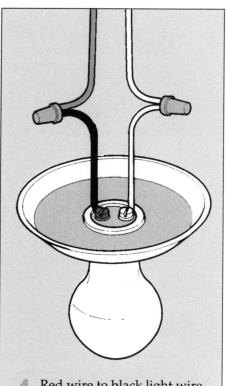

4 Red wire to black light wire, if prewired; if not, to brass light terminal. White to white. Wire connector, tape any splices.

Single-Pole Switch Controls Fixture & Outlet;
Power from Fixture

In this hookup, power runs through the light fixture to a switch and then to an outlet. This circuit requires three-wire cable with ground and two-wire cable with ground, plus pigtail wire.

1 Wiring the Ceiling Box. Power is supplied by a two-wire cable with ground into the box at the light. Splice the black wire from the power source to the red wire of the three-wire cable.

Splice the white wire to the white wire of the three-wire cable and hook the black wire to the black wire of the light fixture. Also, splice the white wire of the power source cable to the light. Use wire connectors and tape the connectors.

2 Wiring the Light Fixture. The black wire connected to the red wire of the three-wire cable is fastened to the brass terminal of the light. The white wire, connected to the power source and the white wire of the three-wire cable, is fastened to the silver terminal of the light fixture. Or the wires are spliced to the white and black wires of a pre-wired light.

3 Wiring the Switch. Fasten the red wire to the top brass terminal of the switch. Make a pigtail of black-insulated wire and connect it to the bottom brass terminal of the switch. Then connect the pigtail to the black wires in the fixture and receptacle cables. Connect the white wire to the white wire in the cable that goes to the outlet. Use wire connectors. The grounding wire (green) is pigtailed and fastened to each box.

4 Wiring the Outlet. Fasten the black wire to the brass terminal and the white wire to the silver terminal. Break the tab if half of the outlet will be switch-operated. Connect the ground with a pigtail to the box.

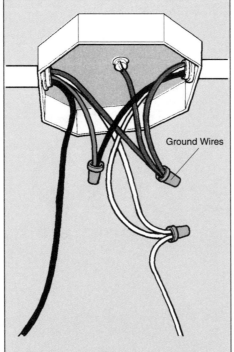

1 The two-wire power cable connects to the three-wire cable from the light to switch. Splice black to red; pigtail the ground wire.

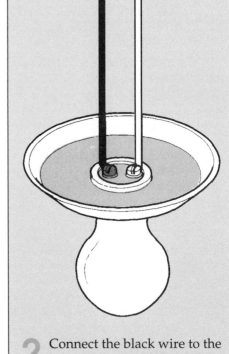

2 Connect the black wire to the brass light terminal and the white wire to the silver or light-colored light terminal.

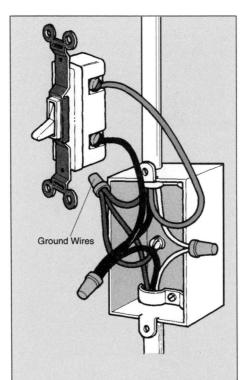

3 At the switch, red hooks to top terminal and black to bottom terminal. White by passes switch; ground is connected.

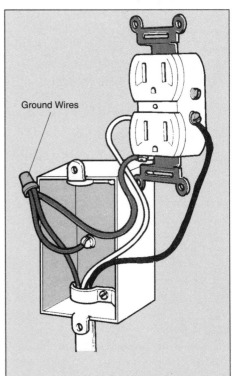

4 Black to brass terminal; white to silver at the outlet. Break metal tab on outlet if half will be switch operated.

Single-Pole Switch Controls Outlet Only

Turn off the power before beginning work and use a two-wire cable with ground throughout.

This project shows how to wire a switch for a light or other device that has been operating from an outlet.

1 Wiring the Outlet. Connect the incoming white wire from the power source to the silver terminal on the outlet. Connect the outgoing black wire to the bottom brass-colored terminal on the outlet. Then connect the incoming black wire to the outgoing white wire. Mark the white wire HOT by wrapping black electrical tape around it. The ground wire is pigtailed to the metal box, to the outlet grounding screw and to the outgoing ground valve. Twist wire connectors around all splices and wrap the joints with electrical tape.

2 Wiring the Switch. Connect the white wire to the top brass-colored terminal. Then wrap this white wire with electrical tape to indicate a hot wire. Connect the black wire to the other switch terminal. Fasten the ground wire to the junction box. If the switch has a ground terminal, pigtail the ground wire and connect it at the box and the switch grounding terminal.

Making Terminal Connections

To make terminal connections, you will need approximately 6 inches of wire in the box. Strip about 1/2 to 3/4 inch of insulation from the ends of the wires without nicking the metal conductor. Use a pair of needlenose pliers to bend each wire end into a hook around the terminal screw with the opening to the right. Tighten each screw, securing its wire. Fold the extra wire like an accordion and place the switch or receptacle in its box. The wire is stiff but use your fingers, not pliers, to avoid damaging it.

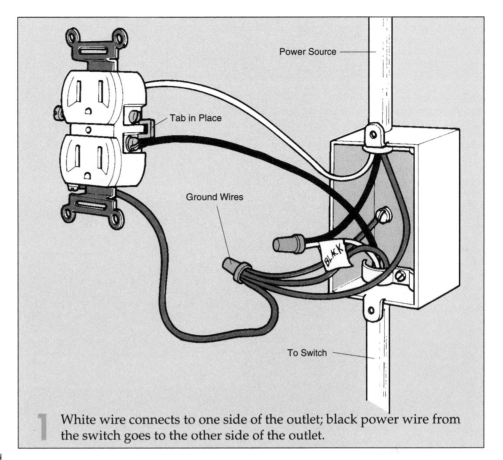

1 White wire connects to one side of the outlet; black power wire from the switch goes to the other side of the outlet.

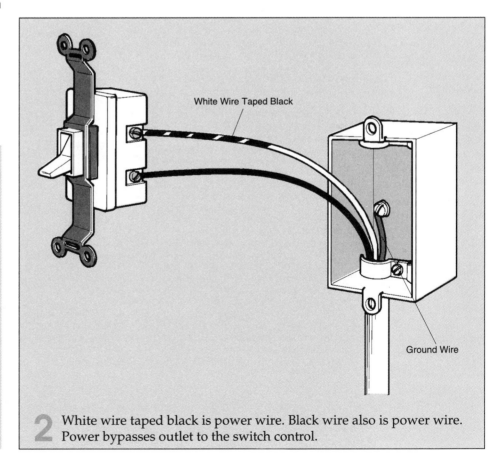

2 White wire taped black is power wire. Black wire also is power wire. Power bypasses outlet to the switch control.

Single-Pole Switch Controls Split Outlet;
Power from Outlet

Use this hookup when you want a single-pole switch to control half of an outlet with the other half of the outlet (bottom) hot at all times. This installation might be in a living or family room where you want to control table lamps along a circuit with a switch, but want other outlets hot at all times.

A two-wire with ground cable is used throughout this circuit with the power coming through the outlet. If this project involves a new circuit, make all the wiring/outlet/switch hookups and then let an electrician connect the circuit to the power supply.

Turn off the power before doing any work on the circuit.

1 Wiring the Outlet. The black wire, using a pigtail, is connected from the power source to the bottom brass terminal of the outlet. The white wire is connected to the upper silver terminal of the outlet.

Wrap the white wire running from the outlet to the switch with electrical tape to indicate that it is now a hot wire. This wire is then spliced to the black wire pigtail and incoming black power wire. The ground is connected, with a pigtail, to the metal box and the grounding terminal of the outlet.

Use a screwdriver or pliers to remove the tab between the brass terminals of the outlet. The switch will then control the upper half of the outlet and the bottom half will be hot at all times.

2 Wiring the Switch. The black power wire is connected to one brass terminal of the single-pole switch. The white wire, coded black with electrical tape, is connected to the other brass terminal. The ground wire is screwed to the metal box. It may be pigtailed and connected to the switch ground terminal if the switch has a ground terminal.

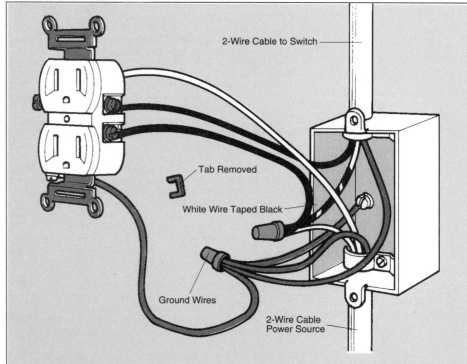

1 A black pigtail connects the incoming black power wire and the black-taped white wire from the switch to one receptacle terminal. The white wire of the incoming power cable connects to the common terminal of the receptacle.

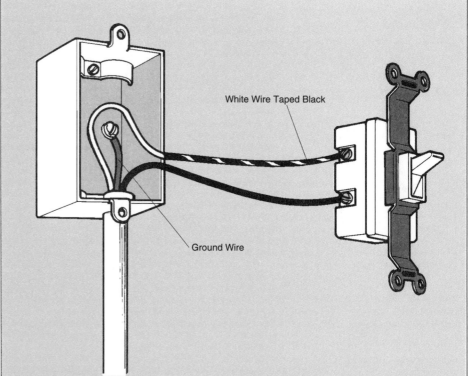

2 The black wire is connected to the switch; the white wire is wrapped black and connected to the switch. The ground wire is connected to the wire box and the switch, if it has a ground terminal.

Adding or Replacing Three-Way Switches

Three-way switches control the power to a light or other electrical device from two separate points. An example is a light in a hallway that can be operated from both the first floor and the second floor. Another example is a light in a garage that can be turned on/off from the garage and also from the kitchen or living room.

Three-way switches require a three-wire system: a power wire and two interconnecting wires called travellers. A fourth, grounding wire also is required except with metal conduit. The proper cable is marked 12/3 WITH GROUND or 14/3 WITH GROUND.

Two three-way switches also are required. Each switch has three terminal screws on the side or back: two on one side, one on the other side. One terminal will be a distinctive color—often black—or will be marked COM, for common. This terminal is for the prime power wire, the black wire in a cable. The other two terminals are for so-called traveller wires that interconnect the switches. When a white wire is used as a traveller in a three-way switch hookup it must be marked with black tape because it too carries power.

Code Requirements

The NEC specifies that all wire must be spliced inside a switch, outlet or junction box. If you splice wire outside the box and there is an electrical fire at this point, your fire insurance coverage could be void.

If you are simply replacing a switch— removing the old switch and installing a new switch— additional wire will not be necessary. In this situation, just connect the new switch to the same wires as the old switch.

Note: You cannot add a three-way switch circuit using two-wire with ground cable.

To add a three-way switch circuit, you will need either a three-wire with ground nonmetallic or BX armored cable, or three wires (black, red, white) to pull through metal conduit. The conduit itself can act as a grounding wire.

How Much Wire?

To figure how much wire you need, measure the distance between the new outlet and the power source. Add an extra foot for every connection you will make along the line. Then, to provide a margin for error, add 20 percent more. For example, if you measure 12 feet of cable between a new and existing receptacle, add another 2 feet for the two connections, making the total 14 feet. Then add 20 percent, about 3 feet, to the total. To do this job you should buy 17 feet of cable.

As mentioned above, wire may not be spliced outside a box. Inside a box, the wire must be spliced together using a twisted wire splice covered by a wire connector and electrical tape. The wire may be attached to a fixture, switch and outlet terminals. To make connections, pull the wire through boxes about 6 inches. Then cut the wire and strip the insulation from the end (see page 18).

Wiring Three-Way Switches

On the following pages, you will find wiring diagrams for three-way switches. By following the paths of individual wires carefully, you can make the connections properly.

Whether you are installing a new circuit or are adding three-way switches to an existing circuit, be sure to identify which wire brings the power into each switch box. It must go to the common terminal of the switch. This is the key to wiring the switches correctly.

When adding new circuits with three-way switches, you should install the wiring for the project and then have a professional electrician connect the circuit to the main service panel.

Solving the Puzzle

In wiring three-way switches you will use two-wire (black and white) and three-wire (black, white, red) cables with ground to make connections between two switches and one or more fixtures, all in individual boxes. Electricians use the procedure described here to make the work go faster.

■ Run lengths of cable from box to box in the circuit. Add enough to make the connections, as explained above.

■ If the power source cable comes into a switch box, connect its black wire to the common terminal of the switch there. If the power cable comes into the fixture box, connect its black wire to the black wire running to one of the switches and connect that to the common terminal.

■ The power cable white wire must connect to the silver terminal of the fixture. If the power comes into the fixture box, connect it directly. If it comes into a switch box, connect it to the white wire of the other cable there. Depending on the hookup, that may go to the fixture box, where you can connect it. If it goes to the other switch box, connect it to the white wire there that goes to the fixture.

■ Connect the common terminal of the second switch to the black wire that goes to the fixture box, and there connect to the brass fixture terminal.

■ Two unconnected wires remain at each switch, red and black or red and white, depending on the layout. Connect these traveller wires to the two open terminals on each switch. If one wire is white, tape both ends black to mark it as a hot wire. If the travellers pass through the fixture box, connect them there: red to red, and black to black (or taped whites together).

■ Where there are two or more grounding wires, connect them with a pigtail to the ground terminal in the box. Where there is only one grounding wire, connect it to the box terminal.

Two Switches Control One Fixture;
Power from Switch

In this circuit, the power cable comes into the first switch box. The path goes through the second switch and on to the fixture.

To install this circuit, you will need three-wire cable with ground between the two switches, and two-wire cable with ground between the second switch and the fixture. Local codes may require the use of conduit, especially for an outdoor light.

Turn off the power to this circuit at the service panel before starting work.

1 **Wiring No. 1 Switch.** Power enters the first switch box on a two-wire cable with ground. Hook the black or power wire to the common terminal on the switch. Connect the white wire to the white wire of the three-wire cable going to switch No. 2. Connect the red and black wires in the three-wire cable to the two lower terminals on switch No. 1. Connect the grounding wires in both cables to a pigtail connected to the box ground terminal.

2 **Wiring No. 2 Switch.** Connect the black and red wires in the three-wire cable from switch No. 1 to the two lower terminals of the switch. Connect the white wire to the white wire of the two-wire cable that goes to the light. Connect the black wire in the light cable to the common terminal of switch No. 2. Connect the cable grounding wires to a pigtail attached to the box.

3 **Wiring the Fixture.** Connect the black wire in the two-wire cable from switch No. 2 to the black lead or brass terminal of the fixture. Connect the cable white wire to the white fixture lead or silver terminal. Connect the cable grounding wire to the box grounding terminal.

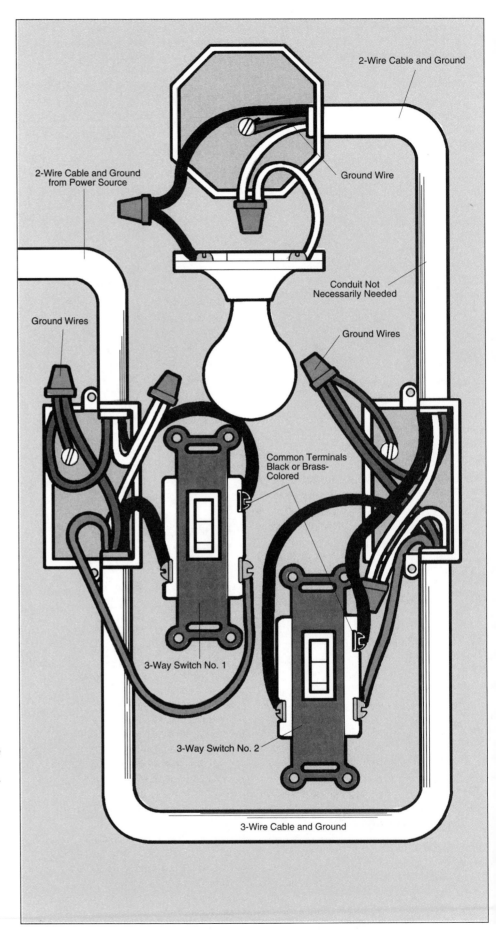

2-Wire Cable and Ground

2-Wire Cable and Ground from Power Source

Ground Wire

Conduit Not Necessarily Needed

Ground Wires

Ground Wires

Common Terminals Black or Brass-Colored

3-Way Switch No. 1

3-Way Switch No. 2

3-Wire Cable and Ground

Two Switches Control One Fixture;
Power from Fixture

In this setup, the power comes into the light fixture on a two-wire cable with ground. The power is wired to pass through the fixture box to the two switches and then return to the fixture. A two-wire cable with ground is used between the fixture and one switch, and a three-wire cable with ground between the two switches. In this circuit the white wire in both cables becomes a hot wire. Therefore it must be marked with black tape.

Turn off the power at the service panel before starting work.

1 **Wiring No. 1 Switch.** Connect the black wire from the cable between switches to the common terminal. Tape the white wire black and connect it to one lower terminal. Connect the red wire to the other terminal. Connect the grounding wire to the box terminal.

2 **Wiring No. 2 Switch.** Connect the red wire from the cable between switches to one lower terminal. Tape the white wire in this cable black and connect it to the other lower terminal. Tape the white wire in the fixture cable black and connect it to the black wire in the switch cable. Connect the black wire in the fixture cable to the common switch terminal. Connect the cable grounding wires to a pigtail to the box ground terminal.

3 **Wiring the Fixture.** Connect the black wire in the cable from switch No. 2 to the black power wire. Tape the white wire in the switch cable black and connect it to the black fixture wire or brass terminal. Connect the white wire in the power cable to the white fixture wire or silver terminal. Connect the cable grounding wires to a pigtail to the box terminal.

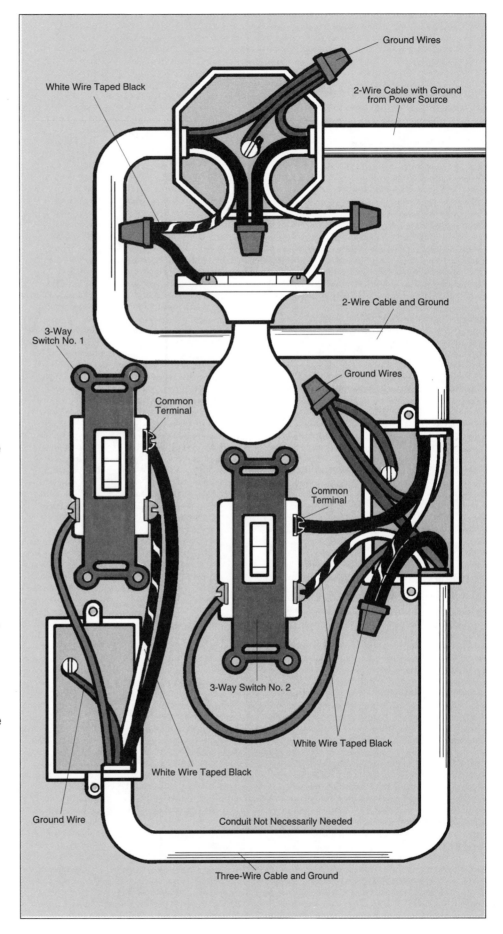

Ground Wires

White Wire Taped Black

2-Wire Cable with Ground from Power Source

2-Wire Cable and Ground

3-Way Switch No. 1

Ground Wires

Common Terminal

Common Terminal

3-Way Switch No. 2

White Wire Taped Black

White Wire Taped Black

Ground Wire

Conduit Not Necessarily Needed

Three-Wire Cable and Ground

Fixture Between Three-Way Switches;
Power from Switch

Here, a light fixture is between two three-way switches with power coming to the first switch on a two-wire cable with ground. The power passes through the fixture box to the second switch and returns to the fixture. A three-wire cable with ground is used between both switches and the fixture. The cable grounding wire (bare or green) is connected to the box of switch No. 2, and to pigtails in the fixture and switch No. 1 boxes. The white wire in the cable between the fixture and switch No. 2 becomes a hot wire in this circuit, so it must be marked with black tape as illustrated and explained in steps 2 and 3.

1 Wiring No. 1 Switch. Connect the incoming black power wire to the common terminal of the switch. Connect the white wire to white wire of the three-wire cable to fixture box. Connect the red and black wires of that cable to the other two switch terminals. Check ground wire connections.

2 Wiring the Fixture. Connect the red wires of the two switch cables together. Wrap black tape onto the white wire coming from switch No. 2 and connect it to the black wire coming into the fixture from switch No. 1. Connect the white wire from switch No. 1 to the white lead or silver terminal of the fixture. Connect the black wire from switch No. 2 to the black lead or brass terminal of the fixture. Check ground wire connections.

3 Wiring No. 2 Switch. Wrap black tape around the white wire. Connect the incoming black wire to the common terminal. Connect the white wire taped black to the terminal below the common terminal. Connect the red wire to the terminal on the opposite side. Check ground wire connections.

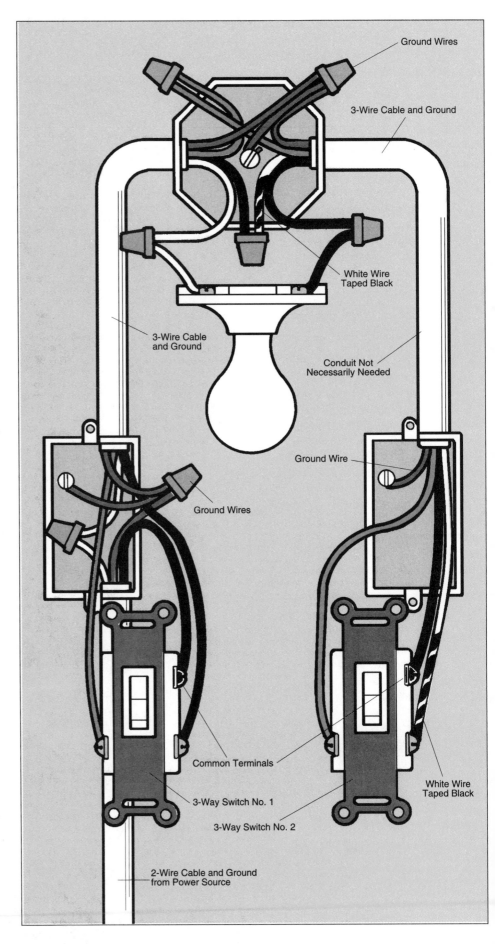

Ground Wires

3-Wire Cable and Ground

White Wire Taped Black

3-Wire Cable and Ground

Conduit Not Necessarily Needed

Ground Wire

Ground Wires

Common Terminals

3-Way Switch No. 1

3-Way Switch No. 2

White Wire Taped Black

2-Wire Cable and Ground from Power Source

Fixture Between Three-Way Switches;
Power from Fixture

In this hookup you can use three-wire cable with ground very easily. The power comes through the light ceiling box. Then you connect it to the switches, which are powered on separate lines from opposite sides of the fixture. Note that the white wire in the power source cable connects directly to the silver terminal of the fixture. The black power wire is connected to the common terminal on switch No. 2. Power is fed back and across to switch No. 1 by white wires coded with black tape to indicate that they are hot between the switches.

1 Wiring No. 1 Switch. The black wire in the cable from the fixture box connects to the common terminal. The white wire is taped black; it and the red wire connect to the other two switch terminals. The grounding wire connects directly to the switch box.

2 Wiring the Fixture. The white wire of the power source cable connects to the white lead or silver terminal of the fixture. The black power wire connects to the black wire of the cable to switch No. 2. The red wires of the switch cables connect together, and the white wires of these cables, taped black, connect together. The black wire from switch No. 1 connects to the black lead or brass terminal of the fixture. The grounding wires all connect to a pigtail attached to the box.

3 Wiring No. 2 Switch. The connections are the same as at switch No. 1. The black power wire goes to the common switch terminal. The white wire is taped black. It and the red wire go to the other two terminals. The grounding wire connects directly to the box.

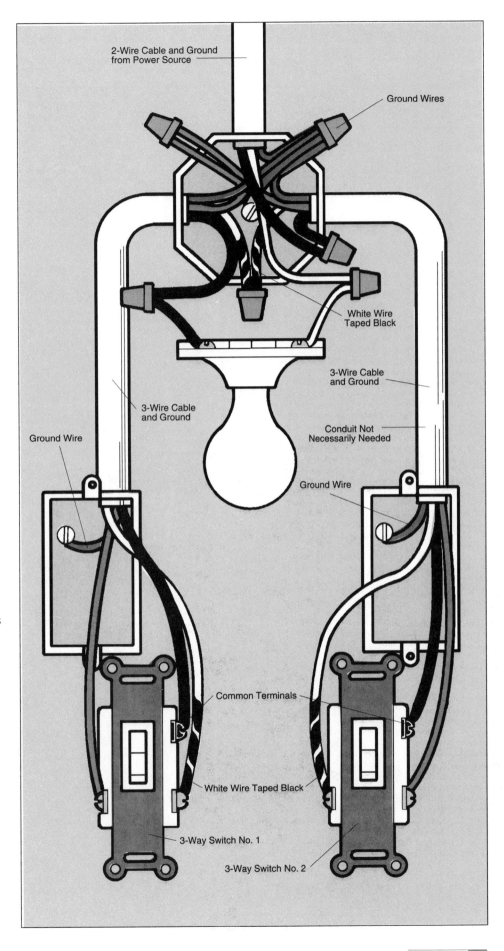

2-Wire Cable and Ground from Power Source

Ground Wires

White Wire Taped Black

3-Wire Cable and Ground

Conduit Not Necessarily Needed

Ground Wire

3-Wire Cable and Ground

Ground Wire

Common Terminals

White Wire Taped Black

3-Way Switch No. 1

3-Way Switch No. 2

Two Three-Way Switches Control Two Fixtures;
Power from Switch

Power comes into switch No. 1 on a two-wire cable with ground. Three-wire and two-wire cables with ground are used between the four boxes.

Note that two white wires specified below must be taped black because they become power-carrying wires.

1 Wiring No. 1 Switch. Connect the black wire in the incoming power cable to the common switch terminal. Connect the white wire to the white wire of the outgoing three-wire cable. Connect the outgoing red and black traveller wires to the other two switch terminals. Pigtail the grounding wires to the box.

2 Wiring the Fixture. In No. 1 fixture box, connect the black from switch No. 1 to the black in cable No. 1 to the next box. Connect the red traveller to the white wire—taped black—in cable No. 1. Connect the whites from switch No. 1 and cable No. 2 to the silver terminal of the fixture. Connect the black wire of cable No. 2 to the brass fixture terminal.

In No. 2 fixture box, tape the white wire in cable No. 1 black and connect it to the red traveller to switch No. 2. Connect the black in cable No. 1 to the white—taped black—going to switch No. 2. Connect the black wires from cable No. 2 and the switch cable to the brass fixture terminal. Connect the white wire in cable No. 2 to the silver fixture terminal.

In both fixture boxes, pigtail all grounding wires to the box terminals.

3 Wiring No. 2 Switch. Connect the incoming black wire to the common terminal of the switch. Tape the white wire black and connect it to one open terminal; connect the red wire to the remaining terminal. Connect the grounding wire to the box terminal.

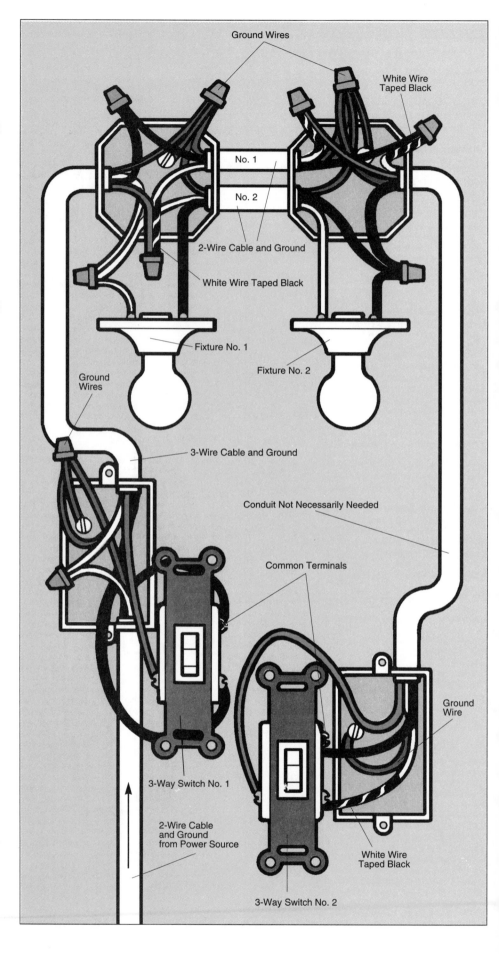

Ground Wires

White Wire Taped Black

No. 1

No. 2

2-Wire Cable and Ground

White Wire Taped Black

Fixture No. 1

Fixture No. 2

Ground Wires

3-Wire Cable and Ground

Conduit Not Necessarily Needed

Common Terminals

Ground Wire

3-Way Switch No. 1

2-Wire Cable and Ground from Power Source

White Wire Taped Black

3-Way Switch No. 2

Two Three-Way Switches Control Two Fixtures;
Power from Fixture

In this arrangement, power comes into one fixture box on a two-wire with ground cable. The two fixture boxes have a three-wire leg between them, as do the switch boxes, but the leg between switch box and fixture box requires only a two-wire cable. The (green) grounding wires are connected to the metal boxes throughout the run.

1 Wiring No. 1 Fixture. Connect the black wire of the power source cable to the black wire in the three-wire leg to the next box. Connect the white power cable wire to the white wire in the ongoing leg and to the silver fixture terminal. Connect the red wire in the ongoing cable to the brass fixture terminal.

2 Wiring No. 2 Fixture. Connect the black wire from the first fixture box to the black wire in the cable to switch No. 1. Connect the white wire from the first box to the silver fixture terminal. Connect the red wire to the white wire—taped black—to the switch, and with a pigtail to the brass fixture terminal.

3 Wiring No. 1 Switch. Connect the black wire coming from the fixture box to the common terminal. Tape the white wire of that cable black and connect it to the black in the three-wire cable that goes to switch No. 2. Tape the white traveller in the cable to switch No. 2 black. Connect it to one open switch terminal and connect the red traveller to the other terminal.

4 Wiring No. 2 Switch. Connect the incoming black wire to the common switch terminal. Tape the white wire black and connect it to one open switch terminal. Connect the red to the other terminal.

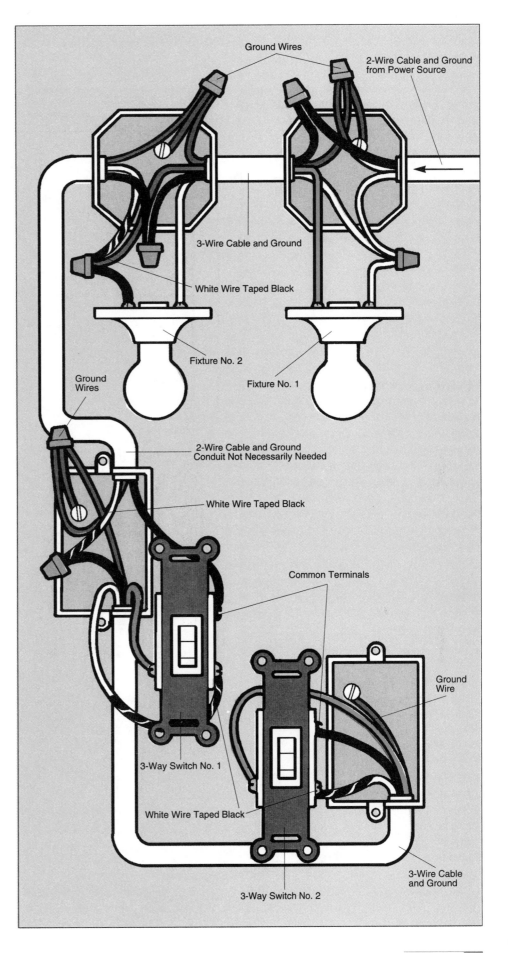

Ground Wires

2-Wire Cable and Ground from Power Source

3-Wire Cable and Ground

White Wire Taped Black

Fixture No. 2

Fixture No. 1

Ground Wires

2-Wire Cable and Ground Conduit Not Necessarily Needed

White Wire Taped Black

Common Terminals

3-Way Switch No. 1

White Wire Taped Black

Ground Wire

3-Way Switch No. 2

3-Wire Cable and Ground

Two Three-Way Switches Control End-of-Run Fixtures

In this hookup, the two lights are at the end of the circuit with the power coming through the first switch, running to a second switch, and then on to the light fixtures.

Since only two-wire cable is needed for the fixture-to-fixture and the fixture-to-switch wiring, you will save money if either or both of these legs in the run is long. Note that in this circuit the red and black wires in the three-wire cable between the switches are the traveller wires.

Throughout the circuit, the grounding wires connect to each other and to pigtails to the metal boxes.

1 **Wiring No. 1 Switch.** Connect the incoming black power wire to the common terminal. Connect the power cable white wire to the outgoing white wire. Connect the outgoing red and black traveller wires to the open switch terminals.

2 **Wiring No. 2 Switch.** Connect the incoming red and black to the two lower switch terminals. Connect the incoming and outgoing whites together. Connect the outgoing black to the common terminal.

3 **Wiring No. 1 Fixture.** Connect the incoming and outgoing black wires together and to the brass fixture terminal. Connect the two white wires together and to the silver fixture terminal.

4 **Wiring No. 2 Fixture.** Connect the incoming black wire to the brass fixture terminal and connect the white wire to the silver terminal. Connect the grounding wire to the box terminal and check that the grounding connections are correct and secure in the other three boxes.

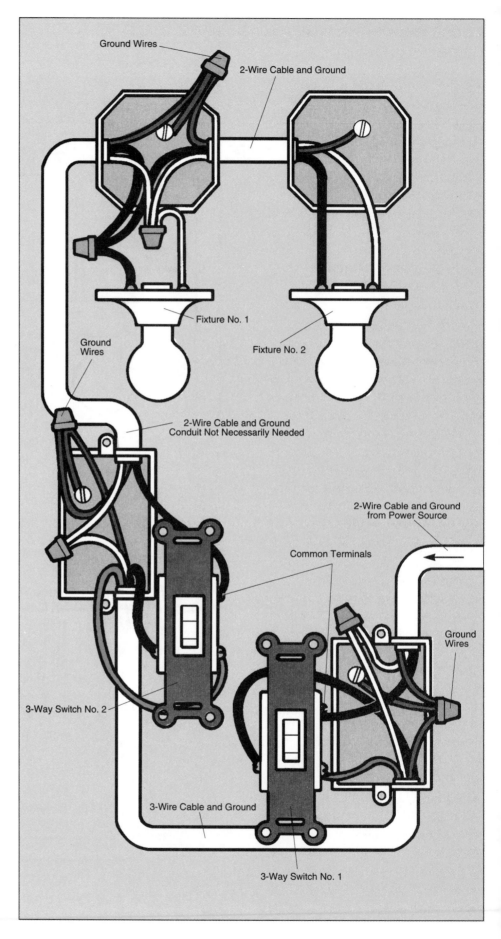

Ground Wires

2-Wire Cable and Ground

Fixture No. 1

Fixture No. 2

Ground Wires

2-Wire Cable and Ground
Conduit Not Necessarily Needed

2-Wire Cable and Ground
from Power Source

Common Terminals

Ground Wires

3-Way Switch No. 2

3-Wire Cable and Ground

3-Way Switch No. 1

End-Wired Switches; Power from Fixture

Power is furnished by a two-wire cable with ground coming into the first fixture box. It is routed to the first switch, then by the traveller wires to the second switch and finally back to the lights. One traveller is red, the other is a white wire marked with black tape. The grounding wires are pigtailed to the metal fixture boxes and connected directly to the switch box terminals.

As in the previous two-light hookups, the switches operate both lights, but the wiring arrangement ensures that even if one bulb should burn out, the other will still work.

1 **Wiring No. 1 Fixture.** Connect the incoming power cable black wire to the black wire going to switch No. 1. Connect the power cable white wire to the white in leg No. 2 to the other fixture and to the silver terminal of the fixture in this box. Connect the black wire in leg No. 2 between fixtures to the brass fixture terminal.

Connect the red wire from switch No. 1 to the black in leg No. 1 to the other fixture. Tape the white from switch No. 1 black and connect it to the black in leg No. 1.

2 **Wiring No. 2 Fixture.** Connect the white in leg No. 2 to the silver fixture terminal. Connect the black in leg No. 2 to the black to switch No. 2 and to the brass fixture terminal. Connect the black in leg No. 1 to the red going to switch No. 2. Tape the whites in the leg No. 1 and switch No. 2 cables black and connect them together.

3 **Wiring the Switches.** Both switches are wired the same way. Connect the incoming black wire to the common terminal. Tape the white wire black and connect it to one lower terminal. Connect the red wire to the other terminal.

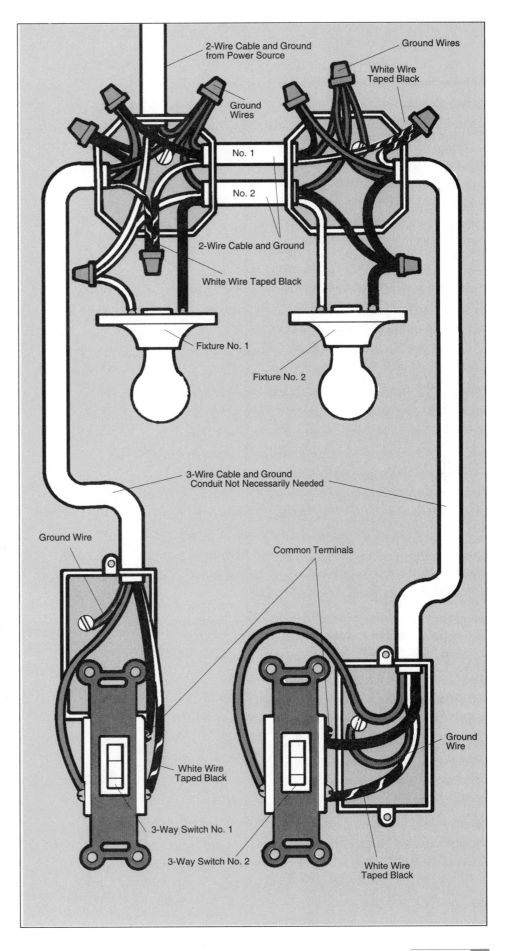

Installing Incandescent Dimmer Switches

A dimmer switch allows you to select different intensities of light to create a mood or to conserve electricity. You must not use them to control receptacles into which you may plug appliances or power tools. This could result in damage to the dimmer switch.

Dimmer switches are available in several styles: toggle-type, rotary-type and slide-type. You can replace any standard single-pole switch with a dimmer. A three-way switch will require a special three-way dimmer.

1 Connecting New Switch. Splice the leads on the switch to the incoming hot wire in the box and to the outgoing wire to the fixture. One of these will be black, the other will be white taped black. Use wire connectors and tape them in place. Then mount the switch in the box.

2 Setting the Dimmer. Turn on the power. Push the dimmer control knob onto the dimmer switch shaft. Then turn the control knob so the light is turned down to its low lighting capacity. Now turn the knurled adjustment nut, located at the base of the shaft on which the control knob fits, counterclockwise with pliers until the lamp flickers. Then turn the nut clockwise until the lamp stops flickering. The adjustment is made.

3 Installing the Cover Plate. When the control knob is set to your satisfaction—high to low light— remove the control knob and mount the faceplate on the switch housing. It screws on just like a conventional faceplate. Then reinstall the control knob. Test the light from high to low. When the knob is on the high setting, the light should be fully on. When it is on low, the light should be at its lowest level. If not, you will have to remove the faceplate and adjust the nut.

1 Join the switch leads to the incoming power and fixture hot wires. Power and fixture white wires join directly.

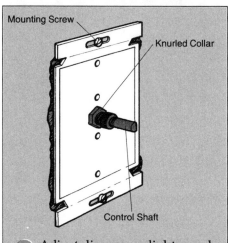

Mounting Screw

Knurled Collar

Control Shaft

2 Adjust dimmer so lights work as you want. Do this by manipulating the knurled nut on the control shaft of switch.

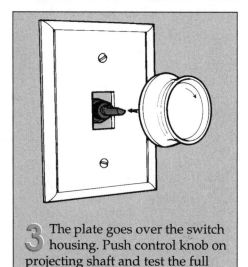

3 The plate goes over the switch housing. Push control knob on projecting shaft and test the full range of the light.

Three-Way Dimmers

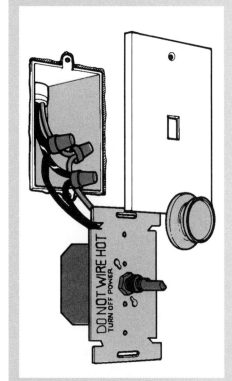

In a three-way circuit, you replace only one of the two conventional switches. To function correctly, a three-way dimmer must be paired with a conventional three-way toggle switch. Turn off the power. Tag the black wire on the common terminal. Then disconnect the old switch. Connect the hot wire of the dimmer switch to the tagged hot wire in the box. Connect the two traveler or switch wires of the dimmer to the two traveler wires in the switch box.

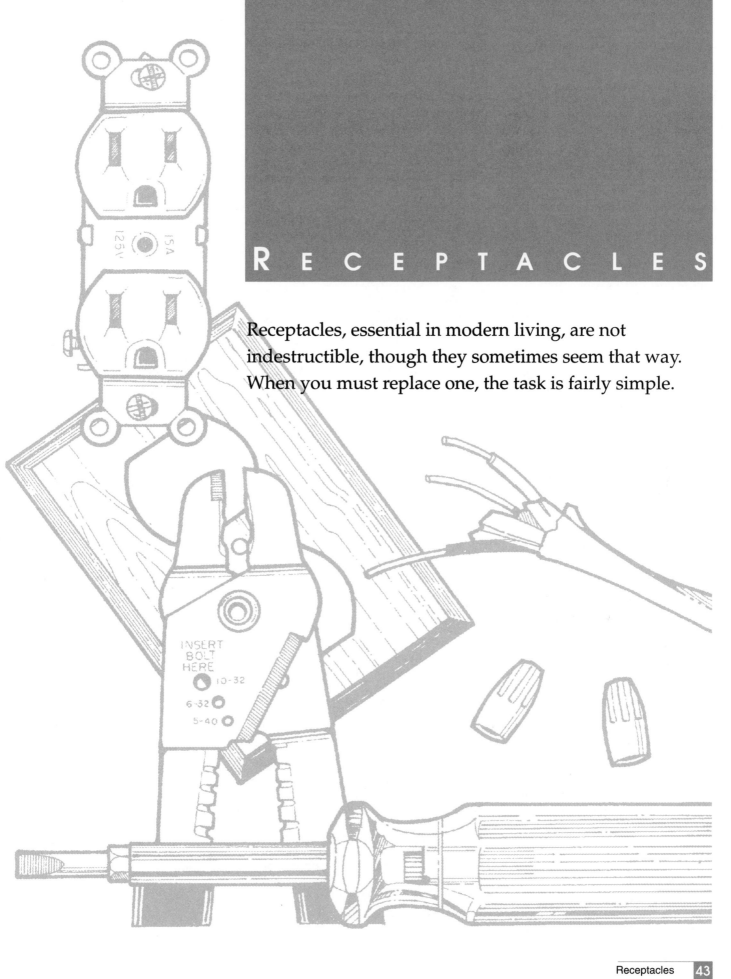

RECEPTACLES

Receptacles, essential in modern living, are not indestructible, though they sometimes seem that way. When you must replace one, the task is fairly simple.

INSERT
BOLT
HERE
10-32
6-32
5-40

Replacing Receptacles

Before you replace a receptacle (outlet), or even remove the faceplate covering it, turn off the power to the outlet circuit. You can check to see if the power is off by plugging in a lamp. Or you can use a tester as described on page 45.

Receptacles are housed in metal or plastic boxes similar to switch boxes. The boxes are covered with a faceplate usually held by a single screw. Behind the faceplate, the receptacles are held by two mounting screws to a metal mounting strap and the box. When these screws are removed, the receptacle may be pulled gently from the box.

To replace a receptacle, you do not have to do anything to the box and you do not have to replace any wiring. You replace only the receptacle.

If the box is tilted a bit left or right in the wall, do not try to straighten it. The wide slots in the receptacle mounting strap will let you shift the receptacle to get it aligned vertically. Then tighten the mounting screws.

The Wiring Layout

Although you do not replace any wires, consider the position of the receptacle in the circuit. This affects the way the receptacle is wired. The box falls either at the middle or at the end of a circuit. Determine the position by the number of cables, or sets of wires, that enter the box through openings in the back or sides.

Each set of wires includes one or two hot wires covered with black or red insulation, which carry live current. If you spot a wire taped with black electrical tape, consider this wire a hot wire.

Each set of wires also includes one with white insulation. Often miscalled "neutral" from earlier wiring practice, the white wire carries power whenever any device in a circuit is operating. It completes the path that must run from service entrance panel to device, back to panel.

If there is an equipment-grounding wire, it will be bare or in green insulation. This wire provides a path to quickly trip the branch circuit breaker in case a hot wire in a piece of grounded equipment comes in contact with the metal equipment case.

"End-of-the-run" wiring has only one set of two or three wires entering the box. The black and white wires attach to the terminal screws. The bare or green wire, the grounding wire, loops around a screw attached to the metal.

"Middle-of-the-run" wiring has two sets of wires entering the box. The hookup varies according to the type of receptacle and the type of ground system used.

Replacement Data

When you buy a replacement (or new) receptacle, be sure you get the one that matches the circuit. The markings and ratings on old and new equipment must match. The markings below are those found on receptacles.

■ Underwriters' Laboratories (UL) and Canadian Standards Association (CSA) monograms indicate that the receptacles have been tested and listed by these organizations. The associations are not connected with the government or special code groups.

■ Amperage and voltage ratings are figures that indicate the maximum amperage and voltage a receptacle can handle. For instance, a rating of 15A-125V means that the receptacle can carry a maximum of 15 amperes of current at no more than 125 volts.

■ The type of current indicated on the receptacle is the only one that the receptacle can use. Receptacles that are used in houses and condominiums in the United States and Canada are marked AC ONLY, which means they are designed for use only with alternating current.

■ Check the types of wire that the receptacle can handle. The wire in your home must match it. The wire in your home will be copper (CO or CU), copper-clad aluminum (CO/ALR), or solid aluminum (ALR). Make sure that the receptacle design can use that wiring.

Receptacle Types

The types of receptacles vary. There are some designed exclusively for use outside; some are made to handle heavy-duty equipment such as major appliances; some are integrated into light fixtures, and some are combined with switches.

The most common home receptacle is the duplex receptacle that is rated at 15 amperes and 125 volts. A duplex receptacle has two outlets and accommodates two pieces of equipment.

Receptacle Wiring

■ Side-wired receptacles, the most common, have two terminal screws on each side. One pair is brass or black in color, the other is silver. A brass terminal always connects to a hot (red or black) wire, a silver terminal only to a white wire. When the break-off link between brass terminals is removed, each terminal will bring power to just one of the two outlets in the receptacle.

■ Back-wired receptacles have openings at the rear in which circuit wires are inserted. Some receptacles have both side- and back-wire terminals.

■ New receptacles also have a green terminal at the bottom to which the equipment grounding wire connects.

Each half of such a receptacle has three openings on the front: two slots for plug blades, and a half-round hole for a grounding prong. In polarized outlets the left slot is wider than the "hot" slot on the right, brass terminal side.

Modern Receptacles

Standard duplex receptacles have two outlets for plugs, as shown at right. Both outlets will be on the same circuit unless the break-off links between terminals on both sides (not shown) are removed—for instance to control one outlet with a switch.

Receptacles are sturdy devices but they can wear out or fail, usually from damage caused by inserting plugs carelessly, with excessive force, and disconnecting devices by yanking on the cord. If plugs fit loosely or fall out, replace the receptacle.

Is the Receptacle Working? Two Tests

The best outlet-testing procedure is to use an inexpensive voltage tester. It has no power, and the test light glows only if the probes connect points where voltage is present.

Another way to test an outlet is with a table lamp. Simply plug the lamp into the outlet and turn the lamp on. If the outlet works, do not fix it. The voltage tester, however, will provide you with a better reading on the outlet and circuit.

1 Testing Power Path. Using a voltage tester, insert a probe into each slot of an outlet. If the wiring is correct and the outlet is working, the bulb will glow.

2 Testing the Ground. To see if the grounding system is working properly, insert one probe in the left outlet slot and the other in the ground-prong hole. The tester should not light. Next check between the ground-prong hole and the right, hot, outlet slot. Now the tester should light. If the outlet fails any test, turn off the power, check the circuit, and replace the receptacle if necessary.

You also can check outlets with an inexpensive three-prong circuit tester that tests for power, grounding and faults such as reversed black and white wire connections.

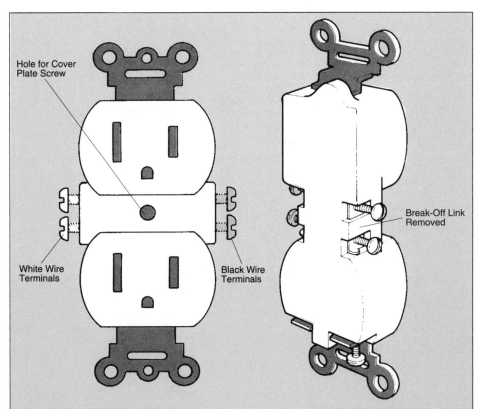

Hole for Cover Plate Screw

White Wire Terminals

Black Wire Terminals

Break-Off Link Removed

Modern Receptacles. A polarized, grounded-duplex receptacle has wide and narrow plug blade slots, half-round ground prong holes and a green terminal for connection to an equipment grounding wire. Usually both outlets are on the same circuit; removing break-off links on each side separates them.

1 Insert one probe of tester in each slot of an outlet. With circuit breaker on, tester bulb should light.

2 With probes as shown, the tester bulb should not light. With one probe in right slot and the other on ground, it should light.

Replacing a Side-Wired Receptacle

1 Removing the Outlet. Turn off the power at the main service panel. Remove the faceplate, which is held by a screw. Remove the two screws holding the outlet in the mounting bracket using a standard slot screwdriver. Then gently pull the outlet from the box far enough to expose the terminals.

2 Noting the Hookup. You will see one or two sets of wires in the box that are connected to the receptacle. Make a note of these connections or tag wires. Black power wires go to the brass terminals; white wires go to the silver terminals; bare or green wires to the ground terminal. Connect a new receptacle that way.

3 Disconnecting/Reconnecting. Remove the wires from the terminals, remove the old receptacle and reconnect the new receptacle. The wires go around the terminals in a clockwise direction. As you tighten terminals, you lock the wires. Then reinstall the outlet in the box.

1 Remove the faceplate and then the mounting bracket screws. Pull out the outlet so the terminals will be exposed.

Ground Wire

2 You will see two or more wires connecting the outlet terminals. Make a note of these connections or tag wires.

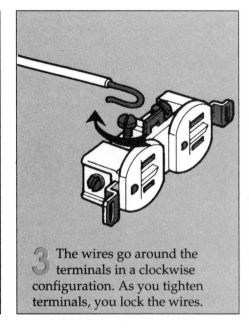

3 The wires go around the terminals in a clockwise configuration. As you tighten terminals, you lock the wires.

Buy the Right Receptacle

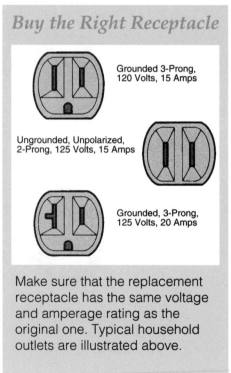

Grounded 3-Prong, 120 Volts, 15 Amps

Ungrounded, Unpolarized, 2-Prong, 125 Volts, 15 Amps

Grounded, 3-Prong, 125 Volts, 20 Amps

Make sure that the replacement receptacle has the same voltage and amperage rating as the original one. Typical household outlets are illustrated above.

End-of-Run Receptacle

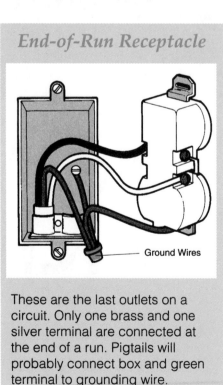

Ground Wires

These are the last outlets on a circuit. Only one brass and one silver terminal are connected at the end of a run. Pigtails will probably connect box and green terminal to grounding wire.

Middle-of-Run Outlets

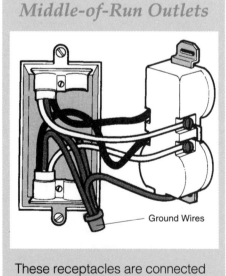

Ground Wires

These receptacles are connected to all wires coming from the box. All grounding wires are spliced together and connected by pigtail to the box ground terminal. If needed, tag wires for reconnection.

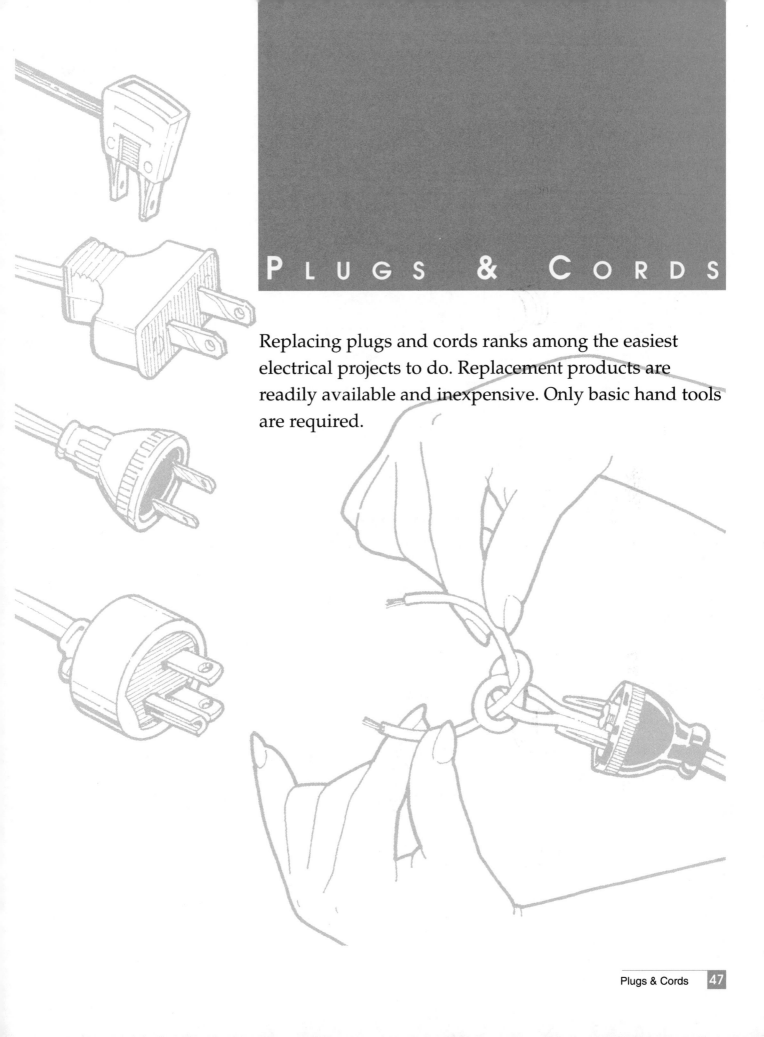

PLUGS & CORDS

Replacing plugs and cords ranks among the easiest electrical projects to do. Replacement products are readily available and inexpensive. Only basic hand tools are required.

Replacing Plugs & Cords

Although it may seem like a complicated job, replacing a plug or a cord or both, is quite simple.

Electrical Plugs

Many lamps and plug-in electrical devices in use today still have a standard-wired or clamp-type plug, neither of which is recognized by the current National Electrical Code. You may find them for sale, however, in electrical departments.

If you are replacing a plug, make sure that the plug meets code requirements. Do not attempt to repair a broken or damaged plug. A replacement is not costly and you are assured that the new plug will perform properly.

Many plugs are permanently attached to electrical cords. That is, you cannot disassemble the plug to disconnect the cord. In this situation, cut the cord immediately behind the plug, strip the insulation and replace the bad plug with a new plug.

Electrical Cords

A variety of line cords is available for lamps and appliances. When replacing a plug, be sure to check the cord for wear and tear. If you see damage, replace the cord along with the plug. Match the cord to the plug and appliance. For example, do not replace heater cord with zip cord.

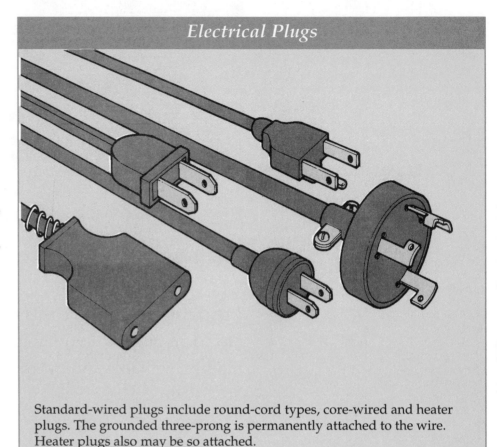

Electrical Plugs

Standard-wired plugs include round-cord types, core-wired and heater plugs. The grounded three-prong is permanently attached to the wire. Heater plugs also may be so attached.

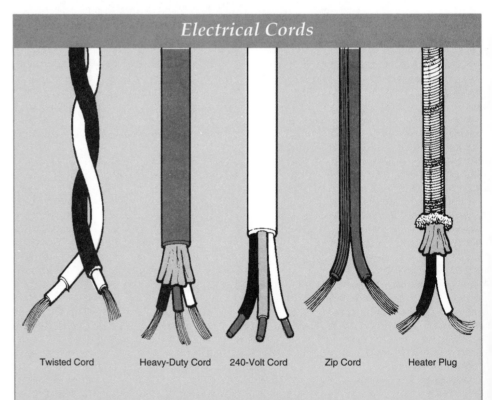

Electrical Cords

Twisted Cord   Heavy-Duty Cord   240-Volt Cord   Zip Cord   Heater Plug

Cords may be precut and packaged in different lengths, or you may buy them by the lineal foot. The third wire in heavy-duty and 240-volt cords is the grounding wire.

Replacing a Standard Plug

Most replacement plugs are wired as shown at right, whatever wire is used; zip cord is shown here.

1 **Removing Old Plug.** With a knife, cut the cord in back of the plug you are replacing. Replace a worn or damaged cord.

2 **Stripping Insulation.** With wire strippers, remove about 3/4 inch of insulation from the wire ends.

3 **Inserting Wire Into Plug.** Thread the cord into the plug. The cord should fit the plug opening tightly.

4 **Tying an Underwriters' Knot.** Split the cord and/or insulated wires inside the cord so you can tie an Underwriters' knot to prevent the cord from pulling loose from terminal screws.

5 **Pulling the Knot Tight.** Pull hard on the ends of the wires to tighten the knot. Then pull the cord down into the base of the plug.

6 **Wiring Around Prongs.** The wire connections go clockwise around the plug prongs and to the terminal screws in the base.

7 **Wiring Around Terminals.** If the wire is stranded, twist it tight and then wrap it around the terminals in the direction the terminal screws turn. Then tighten the terminals.

8 **Installing the Insulator.** Install the cardboard insulator over the prongs and push it down flush.

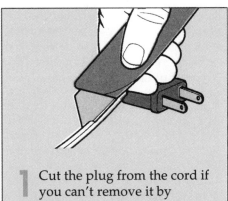

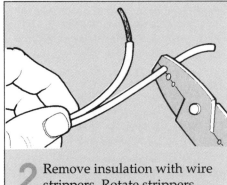

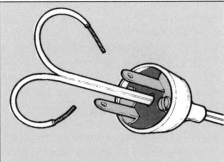

1 Cut the plug from the cord if you can't remove it by loosening the terminal screws.

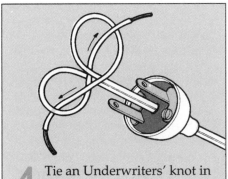

2 Remove insulation with wire strippers. Rotate strippers around wire.

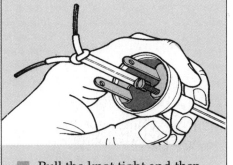

3 Pull the cord through the plug. Cord should fit tightly in the plug opening.

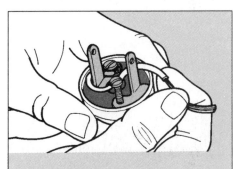

4 Tie an Underwriters' knot in the wires, using this drawing as a guideline.

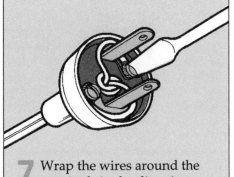

5 Pull the knot tight and then pull the cord down into the base of the plug.

6 Pull the insulated wires around the prongs in the base of the new plug.

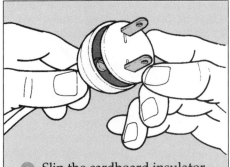

7 Wrap the wires around the terminals in the direction terminals turn. Tighten.

8 Slip the cardboard insulator in place over the prongs. This protects wires.

Replacing a Molded Plastic Plug

Under code provisions, new plugs for home lamps and appliances must be molded in plastic around the cord; they cannot be removed or replaced. Plugs also must be polarized: the prongs are two different shapes that fit a polarized outlet only one way.

If the line cord is damaged only at plug, disconnect cord, cut off the damaged part and attach a new plug. If both plug and cord are damaged, replace both with polarized products.

Heavy-Duty Plug

1. Removable twist-lock plugs are used on some appliances. Loosen the cord clamp, unhook the wires and pull the cord out.

2. Insert the cord and lead the wires around the blades. Hook the black wire to the brass terminal, white wire to the silver terminal and green wire to the green terminal. Tighten the cord clamp.

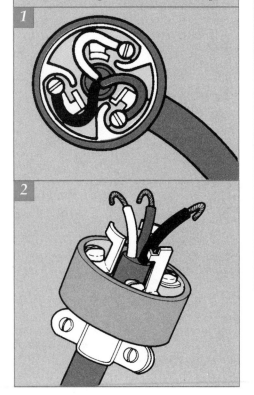

Appliance Plugs

1. You may not be able to find this type of plug on the market. If you can, unscrew the clamshell-like cover and remove wire from terminals.

2. Slip the cord spring onto the cord and into the grooves in the plug base. Then connect the wires to screw terminals. Screw together plug halves.

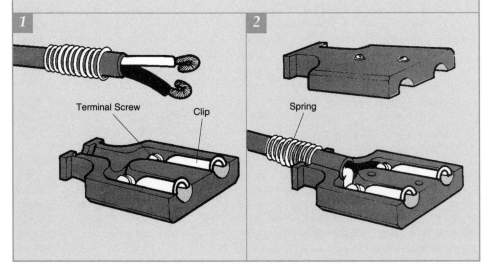

Terminal Screw   Clip   Spring

Flat Core-Wired Plugs

1. Remove the insulator from a flat plug; pull the core out of the housing and remove the wires from terminals. Pull wire through the new plug and separate it.

2. Draw the cord through the housing and separate. Fasten wires to the screw terminals on the core. Seat the core in the housing and replace the insulator.

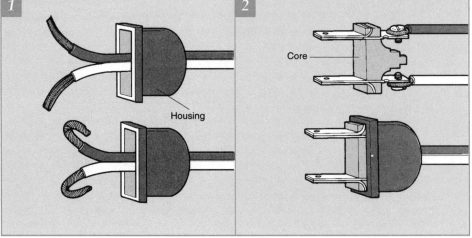

Housing   Core

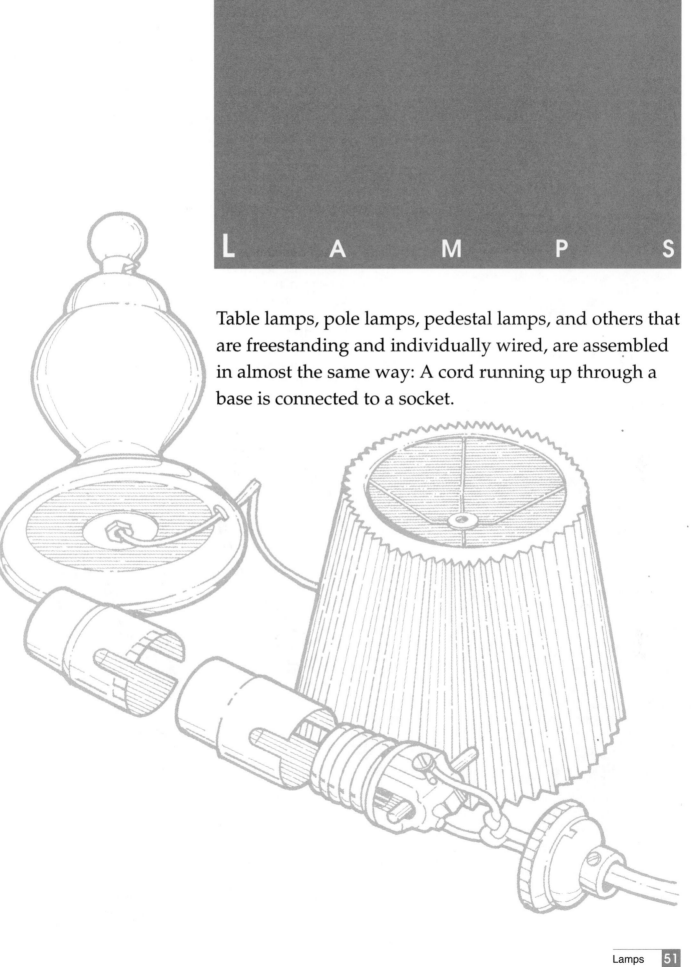

L A M P S

Table lamps, pole lamps, pedestal lamps, and others that are freestanding and individually wired, are assembled in almost the same way: A cord running up through a base is connected to a socket.

In the typical table lamp the cord runs through the base of lamp, sometimes completely through a metal tube and attaches to terminals located in the socket. The light bulb creates heat within the socket, eventually causing the socket to malfunction and to require replacement. If the switch or the plug end of a cord receives a lot of wear-and-tear, they too will require replacement.

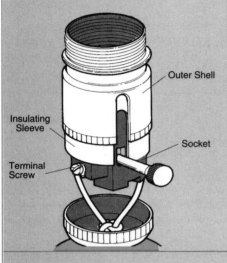

Outer Shell

Insulating Sleeve

Socket

Terminal Screw

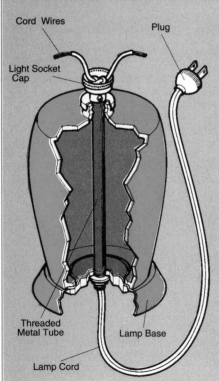

Cord Wires

Plug

Light Socket Cap

Threaded Metal Tube

Lamp Base

Lamp Cord

Rewiring a Lamp

The materials that you will need for this project include a length of cord, sockets, switches, glue, and plastic electrical tape. The tools that are needed include a screwdriver, pliers, wire strippers and a sharp utility knife, pocketknife or paring knife.

Before beginning work, unplug the lamp and place it on a workbench or table surface and have the materials and tools handy.

1 Removing Harp Assembly.
Remove the lamp shade, which is held by a decorative nut either at the top of the shade or where the shade joins the body of the lamp. If the shade is supported by a harp, slide up the small finger nuts at the base of the harp bracket with one hand and lift off the harp with the other.

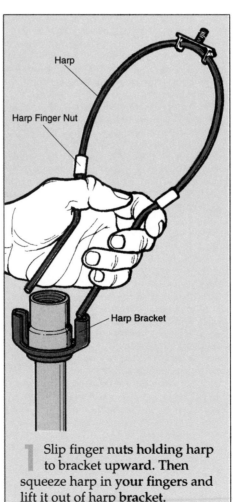

Harp

Harp Finger Nut

Harp Bracket

1 Slip finger nuts holding harp to bracket upward. Then squeeze harp in your fingers and lift it out of harp bracket.

2 Removing Base Covering.
Most table lamps have a felt pad covering the bottom of the lamp. The covering usually is attached with glue. Remove it by breaking the glue seal with the blade of a utility knife, and then peel back the covering so you have access to the cord and/or switch in the base of the lamp.

Most pole and pedestal lamps do not have this covering; the cord runs directly into the lamp housing.

If the lamp has a flat nut and washer assembly that holds the lamp cord secure, turn it counterclockwise and remove it. Most pole lamps have a cord locking nut located just under the shade, remove this in the same fashion.

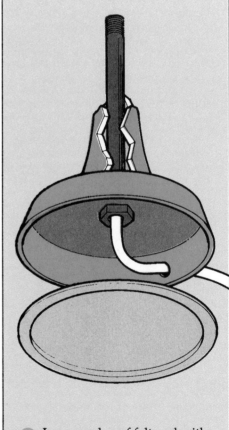

2 Loosen edge of felt pad with a knife and then peel back the pad. This will reveal the locking nut which secures the tube.

3 Removing Lamp Wires.
Depending upon the lamp design there are different ways of removing the lamp wires.

■ **Single Socket Lamps.** Lamp sockets have a thin, brass housing. To remove the outer shell, squeeze in the sides, just above the base cap. (Pry with a thin blade if necessary.) Pull the shell up to expose a cardboard insulating sleeve. Remove the sleeve, loosen the terminal screws and unhook the wires. The copper lamp cord wire always goes to the brass socket terminal, or to a switch lead. The silver wire goes to the silver terminal.

■ **Multiple Socket Lamps.** The wires are connected to multiple sockets the same way single socket wires are connected. However, to replace the cord, you also will have to disconnect the wiring connections in a storage housing near the sockets. The connections are spliced with wire connectors; remove these connectors after you note which wires are connected to each other. You can identify them by the color of the insulation.

■ **Single Switch Lamps.** The lights may not be controlled by switches on the sockets. Instead, a single switch controls the lights. In order to remove the sockets and the cord, you will have to disassemble the connections, which are made with wire connectors. Make a careful diagram of how the wires are joined.

■ **Base Switch Lamps.** If the lamp has a base switch that controls the light sockets, the wiring connections will be located in the base. Make a wiring diagram.

On pole lamps, the lights may be controlled at the socket. Or a single switch may turn on all lights at the same time. In this case, you will have to disconnect the switch assembly in order to remove the cord. Note the connection. Usually it consists of black wires connected to the switch with the white wire bypassing it.

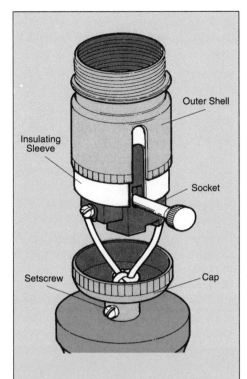

Single Socket. Remove brass socket housing from cap to expose socket terminals. Then remove wires by loosening terminals.

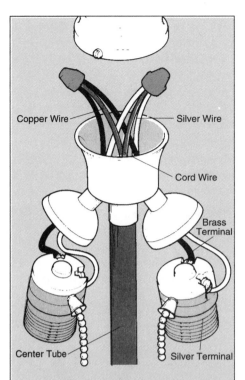

Multiple Socket. For lamps with two or more sockets, remove brass housing and wires from terminals. Then remove switch wires.

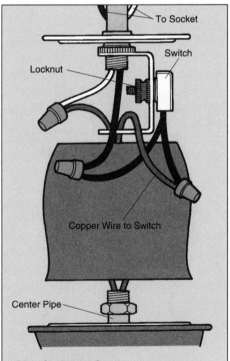

Single Switch. A single switch may control two or more lights. It will be below the sockets. Diagram the hookup as you disconnect wires.

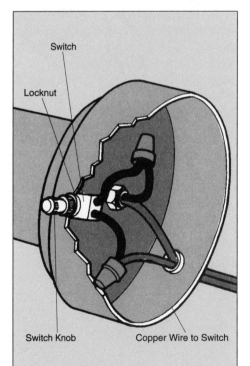

Base Switch. If lamp has a base switch, disconnect the wires here and at the socket(s). Note wiring plan for hookup later on.

4 Removing the Cord.

Temporarily attach the new cord to the old cord at the socket connection. Strip a bit of insulation from the new cord to make a tight joint, and then wrap the joint with a couple of layers of electrical tape so the connection will not pull apart.

Untie any knots or loosen any setscrews in or around the cord on its route through the lamp.

Then carefully pull the old cord out of the lamp base while threading the new cord into the lamp at the same time. When the new cord appears; unwrap the tape, disconnect the old cord and discard it.

Soldered Connections

When you take a lamp apart you may find solder connections between wires, and brackets or tabs. The best way to disconnect these is to heat the solder, not the wire, with a soldering gun until the wire can be pulled free. When you reassemble the connection, you may be able to reheat the solder the same way and reuse it. If not, heat the old solder, scrape it off, and then use rosin core solder (noncorrosive) for new connections.

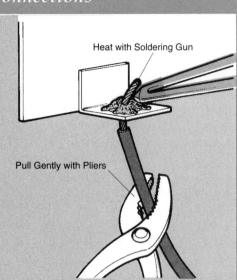

Heat with Soldering Gun

Pull Gently with Pliers

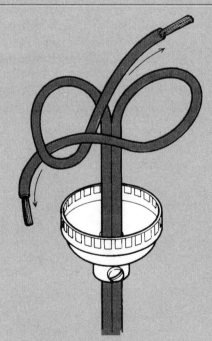

Temporary Splice

4 Splice the new wire to the old wire. Then unscrew and pull both through the lamp at the same time.

Underwriters' Knot

At the socket end, split the insulation (if it is zip cord) and tie an Underwriters' knot or UL knot (Underwriters' Laboratories) in the cord, leaving approximately 3 inches of loose wire at the end of the knot. The knot prevents the wire from pulling loose from terminals in the socket.

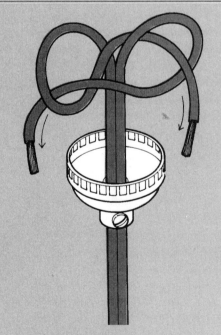

Split the wire and make two fairly large loops. Thread end on one wire through opposite loop.

Thread other wire through the opposite loop, forming loose knot. Pull wire ends apart, making a tight knot.

5 **Connecting the Plug.** It is a good idea to install a new plug on the new wire. Strip about 3/4 inch insulation from the wire, slip on the new plug and tie an Underwriters' knot in the wire.

Pull the wire into the base of the plug and connect the bare wires to the terminals. Then rewire the brass socket (see pages 47-50).

6 **Connecting the Socket.** Strip about 3/4 inch of insulation from each wire. With your fingers, twist the stranded wire as tightly as possible. Then wrap the bare wire around the socket terminals in the direction the terminal screws turn—clockwise.

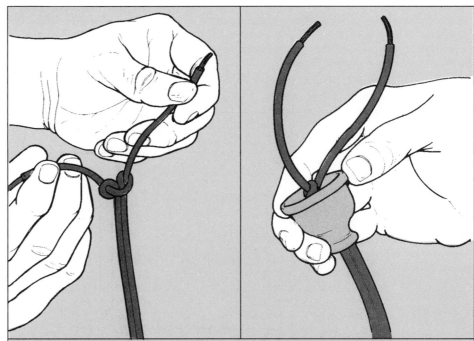

5 Install a new plug on the wire. Strip off 3/4 in. insulation for terminal connections. Then tie a UL knot in wire (left). Pull the knot into the base of the plug. Wrap insulated wire around prongs; screw bare ends to terminals (right).

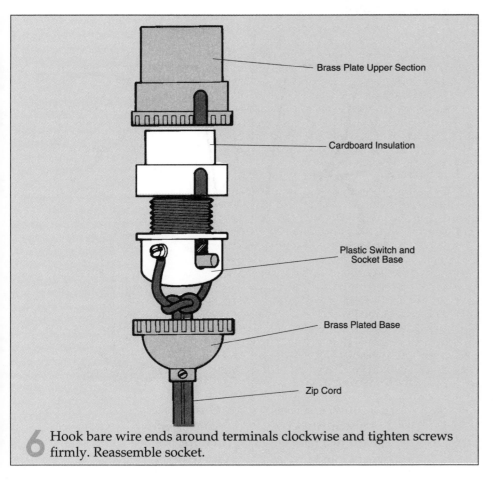

- Brass Plate Upper Section
- Cardboard Insulation
- Plastic Switch and Socket Base
- Brass Plated Base
- Zip Cord

6 Hook bare wire ends around terminals clockwise and tighten screws firmly. Reassemble socket.

Polarized Plug

If your home has polarized receptacles, replace the lamp cord with a polarized plug and cord set. Buy one foot more cord than the total you need. The cord should be the same gauge as the cord already in use: usually No. 18. The large prong goes to the silver wire; small prong goes to hot copper wire.

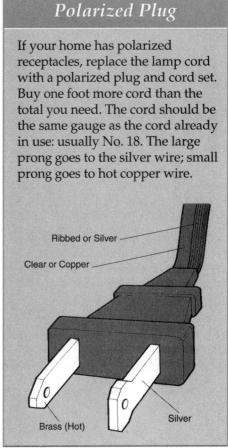

- Ribbed or Silver
- Clear or Copper
- Brass (Hot)
- Silver

Replacing Multisocket Lamp Switches

Two-socket lamps usually have one on/off switch that controls both sockets. Three-socket lamps may have individual switches or a single on-off switch.

Soldered Two-Socket

In this type of lamp, the sockets are molded together. Wire connections run internally between the sockets. The switch turns both sockets on or off at the same time. If one socket will not light, undo the wire connectors holding the switch and the power wires together. Detach the switch wires from the power wires. Connect a new switch.

Separate Sockets

In this two-socket lamp, the sockets are wired separately and can be replaced individually. The switch turns both bulbs on or off at the same time. With the exception of the jumpers, which connect the socket, the replacement is the same as above.

Sockets with Switches

Begin repairing three-socket lamps with individual switches by removing the wires from the socket's terminal screws. Then withdraw the old socket and reattach the wires to the terminal screws of the new socket. Since three sockets receive current through one power cord, there are jumper wires from the line cord to the terminal screws of each socket.

One-Switch Sockets

One four-way switch can control a three-socket lamp. The first position turns on socket No. 1 only; the second position turns on sockets No. 2 and 3; the third position turns on all sockets, and the fourth turns all sockets off. A black wire connects the switch to the line cord, a black wire connects the switch to socket No. 3, and a black wire connects the switch to socket No. 1. Remove the switch by disconnecting wire connectors. Then reconnect them to new switch wires.

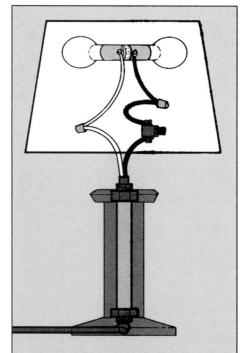

Soldered Two-Socket. To replace a soldered **double socket,** remove wire **connectors and** release splices. **Rewire as shown,** following color **code.**

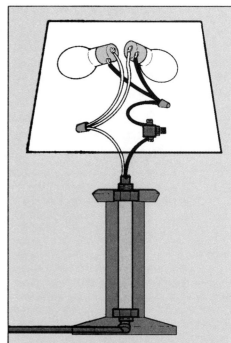

Separate Sockets. If light works independently, replace either **socket separately.** Release wire **connectors;** install the new socket or switch.

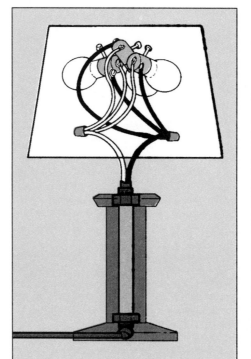

Sockets with Switches. If each socket has its own **switch,** each is replaced separately. **Just release** wire from connectors and rewire with connectors.

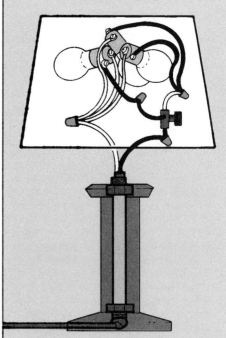

One-Switch Sockets. To replace single-switch on a multisocket fixture, attach a white wire to each socket; then attach a hot wire to each.

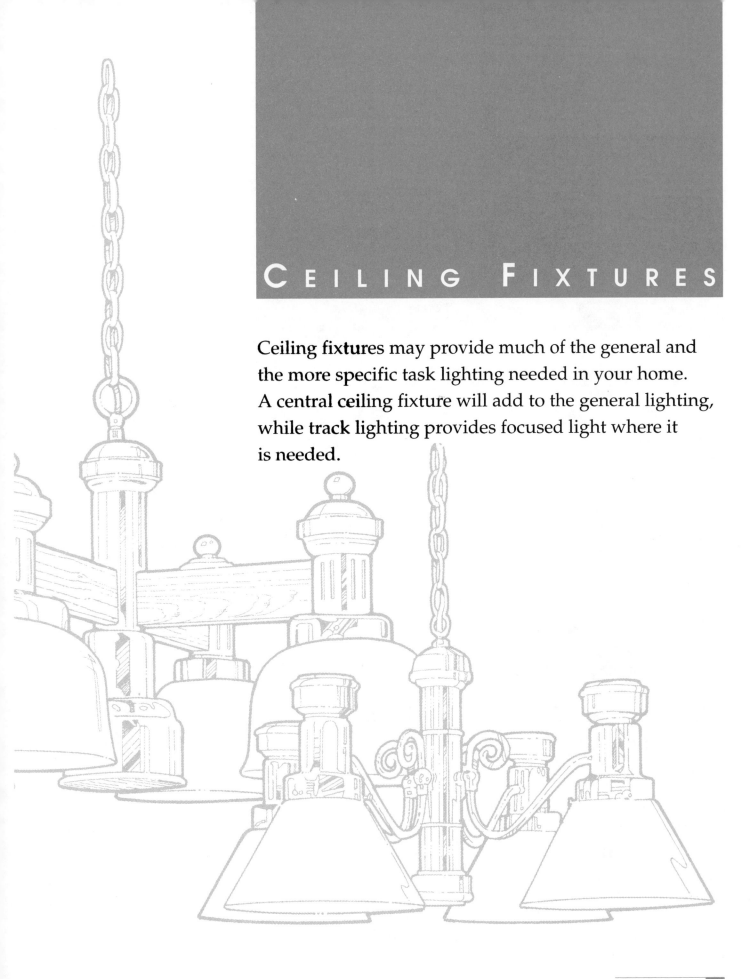

CEILING FIXTURES

Ceiling **fixtures** may provide much of the general and the more specific task lighting needed in your home. A **central ceiling** fixture will add to the general lighting, while **track** lighting provides focused light where it is needed.

Changing a Ceiling Fixture

The job of replacing a ceiling light, such as one over a dining table, with a chandelier is a simple job. Some professionals call this a "change-out."

1 Removing Old Fixture. Turn off the power, then depending on the style of the fixture, remove the globe, light diffuser and bulbs from the fixture. The canopy, escutcheon or fixture base is held to the ceiling electrical box with a locknut or fixture bolts. To remove these fasteners, turn them counterclockwise. This will expose the contents of the ceiling box.

2 Disconnecting Wires. Have a helper hold the fixture while you disconnect the black and white wires from the fixture. If a helper is not available, you can make a hook support from a bent coat hanger to hold up the fixture.

If there are more than two wires in the box, note the configuration and connections. The other wires could be switch and grounding wires.

If the fixture is held by a hickey and nipple or a nut and stud, unscrew these connectors, releasing the fixture.

3 Wiring Fixture. Have a helper hold up the new fixture or support it with a coat hanger arrangement while you connect the fixture wires to the circuit wires. Most fixtures are prewired; remove 3/4-inch insulation from the wires for connection. Mate and twist the black wire of the fixture to the black wire of the circuit; do the same with the white wires and ground wires, if any. Use wire connectors and tape them.

4 Mounting Fixture. The fixture is supported by mounting devices in or on the box, as shown on this page.

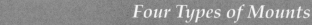

Four Types of Mounts

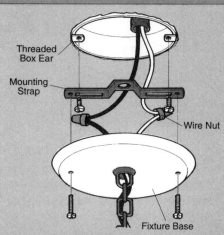

This fixture is held with a mounting strap spanning the ceiling box. Screws hold strap to box and base to strap. Stud is not needed.

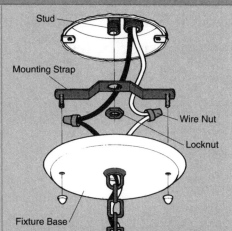

Mounting strap is held to stud with locknut, and fixture base is screwed to strap. Use this arrangement for fixtures weighing less than 10 pounds.

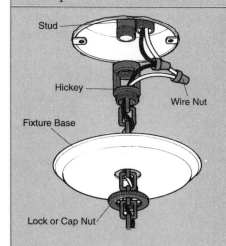

Heavy fixture is mounted on hickey that is screwed to stud in box. Cap nut only secures fixture base to hickey; it doesn't hold fixture up.

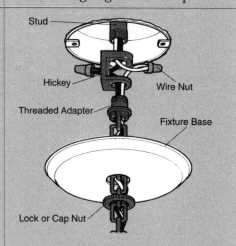

Here, stud, hickey, and threaded adaptor are used to mount very heavy fixture. The necessary parts usually are packaged with the fixture.

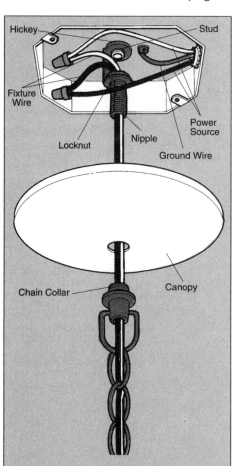

Chandelier-hanging hardware includes a stud, hickey and nipple that supports the extra weight of the fixture. A box knockout accommodates the stud. Wires are spliced black-to-black and white-to-white with wire connectors. Make sure the box is secure to framing.

Track Lighting

Track lighting creates a theatrical mood. It can be formal and informal at the same time, and it can be installed in open ceilings and on finished ones. If you install track lighting in an open ceiling, you can spray the joists and sheathing above flat black and the track lights will seem to float on the ceiling surface. Local lighting with track lights, in many ways, is similar to raceway lighting. The basic part is a length of surface wiring that can be tapped anywhere for a fixture. Because of its flexibility, you can place it in almost any room. The track comes with various adaptors that enable you to add outlets or, in some cases, even a fairly heavy chandelier.

Track lighting installation procedures vary according to the manufacturer and model. Be sure that you read all the instructions carefully before buying the system. You can start with the basic system and add to it as your budget or decorating scheme dictates. Make sure that the brand of track lights you buy will accommodate your plans for the future.

If the track-lighting channel is attached to a ceiling, use toggle fasteners to ensure the stability. If the track is mounted on open framing, you can attach the track to the edges of joists or along a joist edge. In some situations you can recess the track and the lights between the joists in a straight line.

The Power Sources

Track lighting is connected to the house wiring, like any other ceiling fixture. You probably will have to add a ceiling junction box to install this style of raceway, since the existing boxes will not be close enough to the track.

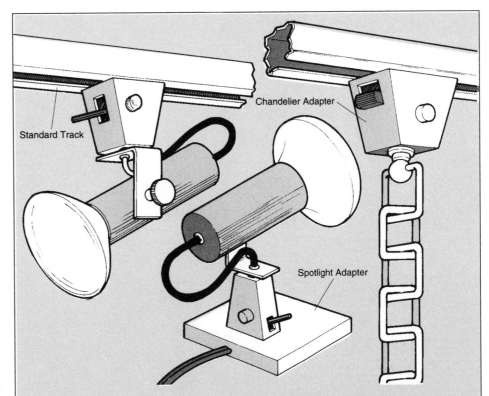

Track Lighting. Track lighting is available with myriad fittings—from standard lights to spotlights to chandeliers. The lights also come in different designs to match most decors.

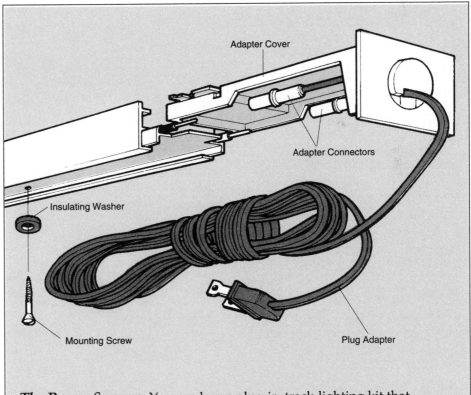

The Power Sources. You can buy a plug-in, track-lighting kit that eliminates the need to make wiring connections. You also can buy track that must be connected to a wired ceiling box or junction box.

Installing Track Lights

Turn off the circuit on which you will be working. Roughly plot the position of the track on the ceiling. Install a ceiling junction box, if none exists, at one end of the track's location and fish cable to the box. If you are using a plug-in track lighting system, install the track as described below and simply plug it into an existing wired outlet.

1 Installing the Connector Plate. The track adaptor plate covers the junction box and holds the track connector and the electrical housing. Assemble these pieces. Splicing like-colored wires together, hook up the track wires to the cable wires. Then fasten the adaptor assembly to the junction box ears with the screws provided.

2 Plotting the Track. Working from the center slot of the track connector, draw a line along the ceiling to pinpoint the location of the track.

3 Installing Track Clips. The track is held in position by special clips spaced evenly along the track. Hold the clips in place on your line, and mark pilot holes in the ceiling. Drill the pilot holes and attach the clips with toggle bolts to a drywall surface or with screws to a wooden surface.

4 Connecting the Channel. Connect the track channel solidly to the electrical connector; slip the channel into the track connector. Snap the track channels into the clips. Then tighten the setscrews along the sides of the clips to hold the channels firmly in position.

To complete the project, install the raceway cover and attach the track lights anywhere you wish.

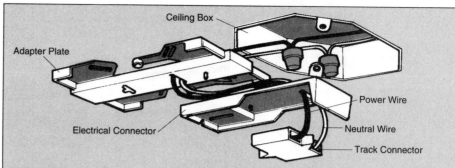

1 Connect the track wiring to the house wiring using the metal adaptor plate. Use wire connectors to connect splices and wrap them with plastic electrical tape. Fasten assembly to ceiling box.

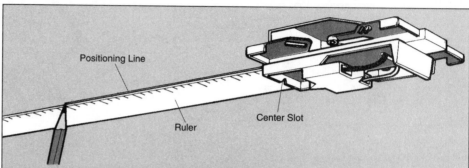

2 To plot the line for the track itself, align a ruler with the center slot on the track connector. Draw the line straight across the ceiling to the location of the opposite end of the track.

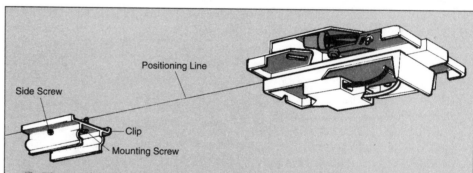

3 The track will be held along the line by plastic clips. Center the clip on the line and draw a mark for the screw hole. Then install the clip using toggle bolts or wood screws.

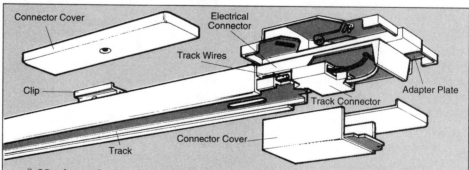

4 Hook up the track to the track connector. The track itself supplies power for the track lights. Make sure that the track connection joints are butted tightly together. Then attach lights.

Fluorescent Fixtures

The three main parts of a fluorescent fixture are the fluorescent tube, which may be straight or circular, the starter and the ballast. Defects in these components cause most fluorescent fixture problems.

The Tube

A fluorescent tube produces light in this way: Inside a tube, the electric current jumps or arcs from a cathode at one end of the tube to an anode at the other end. The tube is filled with mercury and argon gases. As the arc passes through the gases, it causes them to emit invisible ultraviolet light. To make the light visible, the inside of the tube is coated with phosphor powder that glows when hit by ultraviolet light.

The Starter

The starter is a switch that closes when activated by an electric current. After a momentary delay, the starter allows current to energize gases in the tube. There are two types of starters: replaceable ones, which are about 3/4 inch in diameter with two contacts protruding from one end, and a rapid-start fixture.

The starter is built into the ballast and cannot be replaced independently of the ballast.

The Ballast

The ballast is a box-like component usually about 6 to 7 inches long. It is a kind of governor that monitors electrical current so that it is at the level required to provide proper light operation. There are two types of ballasts: Choke ballasts limit the amount of current flowing through the tube. Thermal-protected ballasts, found in fixtures that hold long fluorescent tubes, incorporate transformers and choke coils. When the light is turned on, a transformer steps up the voltage to deliver a momentarily high surge of electricity, causing the tube to glow.

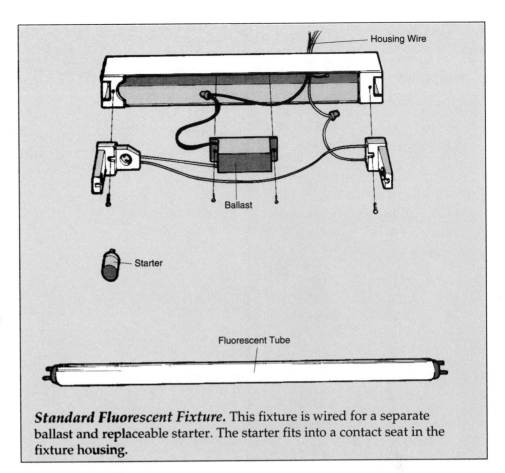

Standard Fluorescent Fixture. This fixture is wired for a separate ballast and replaceable starter. The starter fits into a contact seat in the fixture housing.

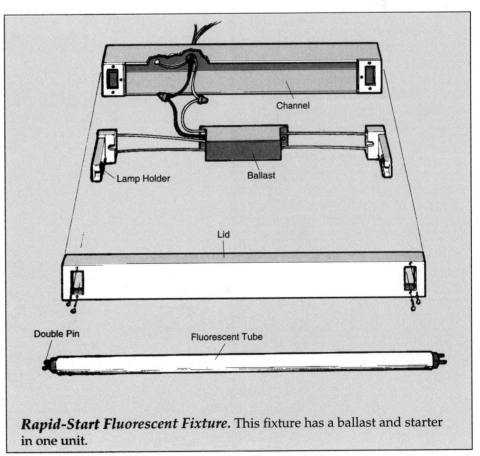

Rapid-Start Fluorescent Fixture. This fixture has a ballast and starter in one unit.

Replacing a Fluorescent Ballast

Turn off the circuit breaker or remove the fuse that supplies power to the circuit. Then remove the tube(s), and take off the cover. Jot down the number codes on the old ballast and take them to the store with you, to make sure you buy the right ballast replacement.

1 Disconnecting the Wires.
Remove the wire connectors or loosen the terminal screws to disconnect the ballast wires. The wire connections of the ballast you are replacing are the same or similar to those shown on this page. Notice that ballast wires are color-coded. A ballast wire of a given color is always connected to the fixture wire of the same color. A thermally protected ballast must be connected in the same way. If the complexities of the wires confuse you, make a simple color-coded diagram before you disconnect any wires.

2 Removing the Ballast.
The ballast you are removing is heavier than you might think. Be careful. Have a helper hold the ballast while you remove the fasteners that attach it to the fixture. Note carefully the alignment of the ballast in the fixture and then take it down.

3 Connecting a New Ballast.
Again, with a helper holding the new ballast, line up the ballast so that it is in the same position that the previous ballast held. Screw the new ballast to the fixture. Match the color codes of the wires and twist these wires together with your fingers or pliers. Then thread the connected wires into wire connectors. Wrap the wire connectors with a couple of layers of plastic electrical tape. As a final step, before you replace the cover, turn on the power and test the light.

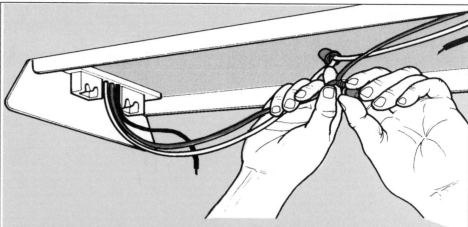

1 When replacing a ballast unit, first make a color-coded diagram of the wiring. Then remove the wire connectors from all connections and disconnect the splices. Note the color-coding of wires and numbers on ballast at the time of purchase.

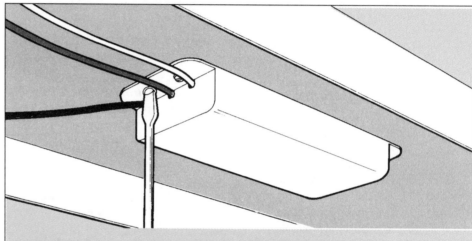

2 Loosen the screws that hold the ballast in the fixture. Have a helper hold the heavy ballast while you do this. Then remove the ballast from the fixture, noting its exact position.

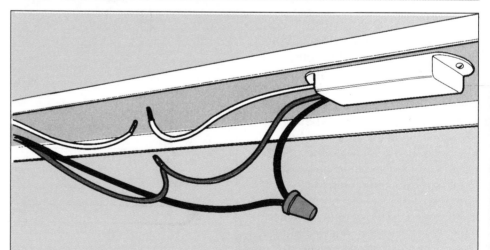

3 Do not worry about the maze of wires. Connections are made by matching colors, splicing wires and securing splices with wire connectors wrapped with plastic electrical tape.

Installing a Fluorescent Fixture

To replace a fluorescent fixture, first turn off the power to the circuit and remove the old fixture from the ceiling.

Circular Fixtures

In the center of the ceiling box, add a threaded stud, if one is not present. The fixture hangs on this stud. Add a reducing nipple to the stud. Have a helper hold the fixture while you connect the power wires: black to black, white to white. Attach wire connectors to the splices and wrap them with electrical tape. Push the wires into the box, thread the nipple through the hole in the center of the fixture, and secure the fixture with a cap nut.

One-Tube Fixtures

You will need a hickey and nipple if the box has a stud. If not, you can attach the fixture to a nipple and strap screwed to the ears in the box. First splice the fixture wires to the house wires, attach wire connectors, and wrap the splices with plastic electrical tape. Then attach the fixture to the ceiling box with the nipple, a washer and a locknut. Have a helper hold the fixture while you assemble and fasten it to the ceiling box. When the fixture is stable, drive a couple of sheet-metal screws through the fixture housing into the ceiling at each end.

Large Fixtures

Fixtures with more than two tubes usually have a center cutout that is used when hanging the fixture from an octagonal box. The fixture uses a stud, hickey, nipple, and a mounting strap inside the housing. The assembly is held with a locknut. Connect the wiring with wire connectors. Then push the wires into the box and secure the fixture.

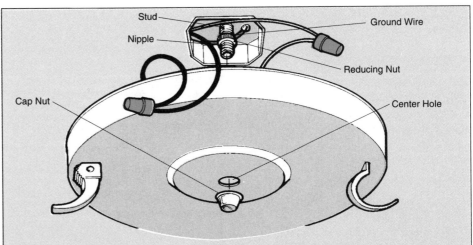

Circular Fixtures. Circular fixture is connected to a stud and nipple inside the ceiling box. Complete the wiring first with wire connectors, then hang fixture with cap nut on nipple. Install the tube.

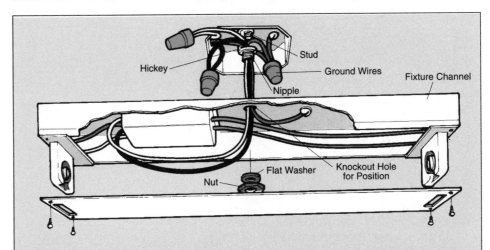

One-Tube Fixtures. Knockouts in housing let you position fixture almost anywhere over box. Punch out knockout. Connect wiring. Fasten to ceiling. Add screws through housing at ends to support fixture.

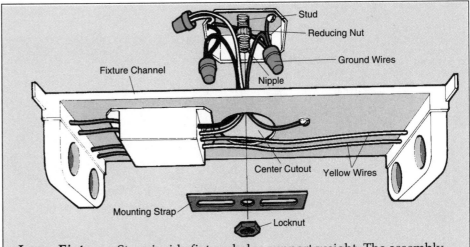

Large Fixtures. Strap inside fixture helps support weight. The assembly order is: stud, reducing nut, nipple, fixture, strap and locknut tightener. Cover plate slips into channels along sides of fixture.

Repairing Fluorescents

Eliminating Flickering

If a tube has the flickers, remove it. Straighten bent pins with pliers. Then burnish the pins lightly with fine-grit abrasive or steel wool.

Replacing Pull-Chain

With pliers, loosen and remove the knurled nut and locknut that hold the switch in the fixture housing. Disconnect the switch wires. If they cannot be disconnected, cut them, leaving a couple of inches for reconnection. Lift out the entire switch. Make wiring connections with wire connectors. Tape the connectors, then set the new switch in place. Secure with locknuts.

Replacing Toggle Switches

The technique for replacing toggle switches is the same as chain switches, except that power wire terminal screws are attached to the switch. Remove the wires, install a new switch, and rewire it.

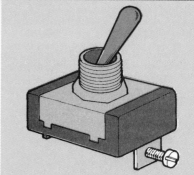

Replacing Toggle Switch. Release the locknuts on the old switch. Attach wires to the terminal screws of the replacement switch.

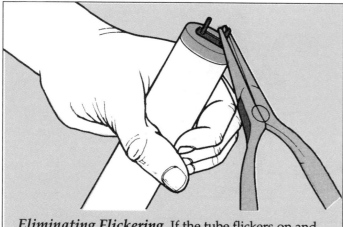

Eliminating Flickering. If the tube flickers on and off, the pins, not the switch, may be the trouble. Straighten the pins and then shine them with abrasive.

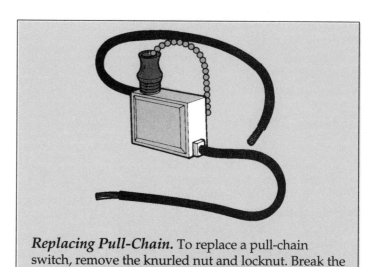

Replacing Pull-Chain. To replace a pull-chain switch, remove the knurled nut and locknut. Break the splices. Put in new switch and reconnect the splices.

Replacing Push Switches

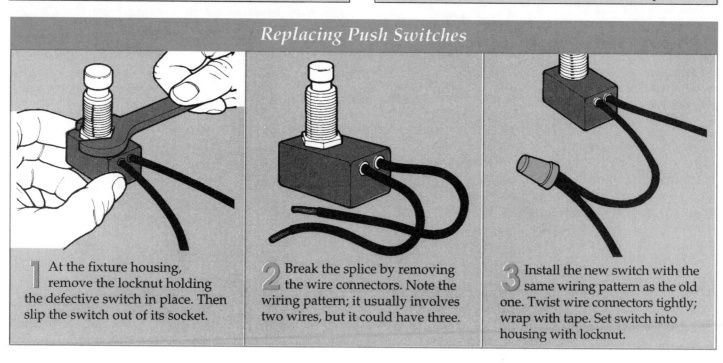

1. At the fixture housing, remove the locknut holding the defective switch in place. Then slip the switch out of its socket.

2. Break the splice by removing the wire connectors. Note the wiring pattern; it usually involves two wires, but it could have three.

3. Install the new switch with the same wiring pattern as the old one. Twist wire connectors tightly; wrap with tape. Set switch into housing with locknut.

Ceiling Fans

The installation of a ceiling fan involves two critical measurements. There must be at least 7 feet of clearance from the floor to the blades of the fan in the room in which the fan will be installed. The blades also must be free to rotate; there can be no obstruction in the path of the rotation. Check the vertical space in the room. Also check the length of fan blades for proper blade clearance.

Ceiling fans operate on regular house power. No electric transformers or special switching devices are needed. You can connect them directly to the wires in a ceiling box listed for this purpose. If you are running cable to a new ceiling box for the installation, the wire size should correspond with the wire size that you are tapping into. If you are creating a new circuit, the wire size should be AWG No. 12/2 with ground.

Switching Choices

Installing a switch also can be considered. If the room in which the fan is installed has a ceiling light controlled by a wall switch, the fan hookup is simple: Remove the light fixture and substitute the fan. If the light fixture has a switch on it, you can buy a fan that is controlled by its own switch.

To control the fan with a wall switch, you have the following options:

■ Tap into a source of power elsewhere in the room, such as an outlet;

■ Tap into the fan; or

■ Run a switch loop from the ceiling box to the wall. If you have an attic crawl space in your home (and you have access to it), you may be able to tap into the wiring in the crawl space. Use either the existing ceiling box, or a different nearby ceiling box. You also could add a junction box from which the ceiling box and fan receive electricity.

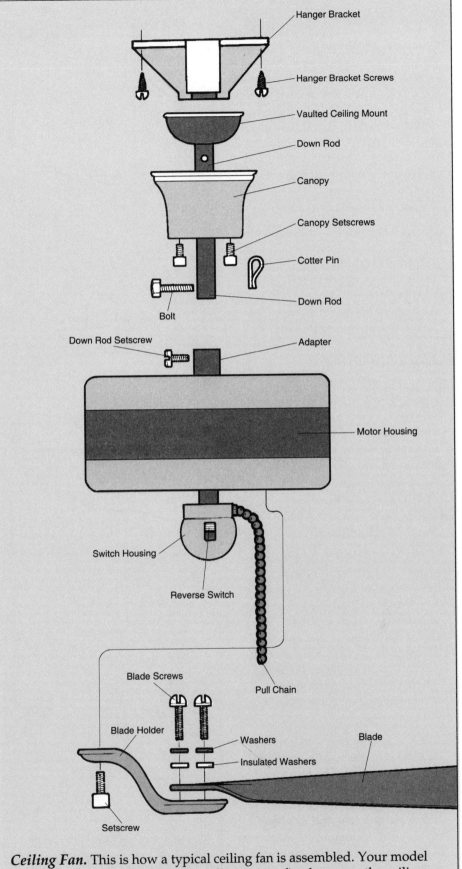

Ceiling Fan. This is how a typical ceiling fan is assembled. Your model may differ slightly. The key to installation is to firmly secure the ceiling electrical box to the joists. The box must be rigid to support the fan.

Brackets must be attached to the framing, and the electrical box must be listed for this specific use. Check the box; it should be securely attached to the joists. If the box feels loose, drive a couple of round-head wood screws through the mounting holes into the framing members.

J Hook (Isolation Type)

An isolation mount is installed on or in the box. It absorbs vibration, torque and sound. The hanger pins must be fastened into the ceiling framing, not into the box alone. Make sure the ceiling box is securely fastened to joists.

Flush Mounting

To install the fan on a finished ceiling where there is no outlet or wiring, you must open the ceiling and add framing between joists to fasten the box or screw directly to framing joist . Do not use an adjustable hanger bar, it may not be strong enough.

Surface Mounting

If you mount the fan on an exposed beam ceiling, position the fan between beams, using a 2x4 or 2x6 length of wood between the beams to hang the fan. You also can use a special hanging bracket, not shown. There must be 6 inches of space for the bracket. It also may be used where attic crawl space is not available for mounting and wiring the fan.

The blocking between beams can be spiked to the beams. Install the box to the beams. The fan is mounted on the box or a J hook also screwed into the box. You can frame the box in wood or other materials to hide it. If you decide to do this, make sure that the box is screwed directly to framing joist.

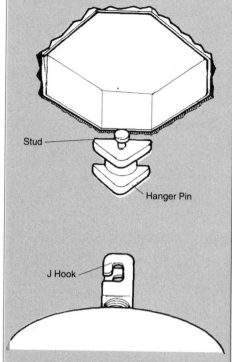

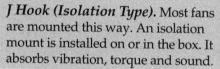

J Hook (Isolation Type). Most fans are mounted this way. An isolation mount is installed on or in the box. It absorbs vibration, torque and sound.

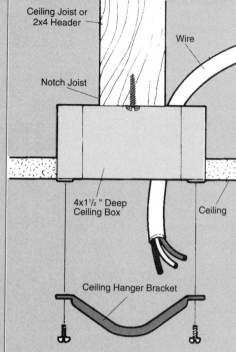

Flush Mounting. Ideally the box for the fan is attached through the ceiling with screws to a framing member.

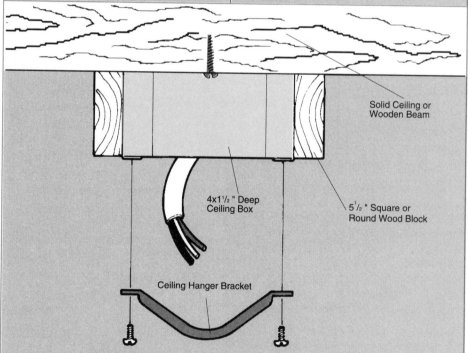

Surface Mounting. If you find it necessary to mount the fan-mounting ceiling box on the surface, you can frame the box in wood or other materials to hide it. If you decide to do this, make sure that the box is screwed directly to a framing member through the ceiling covering.

Installing Ceiling Fans

Most ceiling fans are assembled the same way. There may be slight variations between fan manufacturers, but the differences will be noted in the instructions enclosed in the fan package. Wire colors also may vary.

1 Turning Off Power. Turn off the electric power to the ceiling box at the main electrical service panel. Do not flip a wall switch and assume that the power is turned off. Go to the main service panel and flip the appropriate circuit breaker or remove the fuse. If there is any doubt, use a voltage tester, as described on page 27.

2 Removing Fixture. Remove the ceiling fixture and disconnect the black and white wires from the terminals or wire connectors. If the ceiling light is controlled by a wall switch, the fan also may be controlled by the wall switch. If the light is not controlled by a wall switch and you want to control the fan with a wall switch, you will have to fish cable through the ceiling and down the wall to create a switch loop. If you want to control the fan with the fan switch, simply connect the fan to the power wires inside the ceiling box.

3 Securing Ceiling Box. Be sure the ceiling box is securely mounted to the ceiling framing. Double-check to make sure it is. The best insurance is to drive a couple of screws through holes in the box into solid material. Assemble the mounting bracket according to manufacturer's instructions. Attach the bracket into the ceiling framing, not to the ceiling box alone. Be certain the product conforms to electrical codes.

If the box is mounted on a pitched ceiling or beam, you have to use either a swivel hanger or an angle kit that you can buy at the store. The kits are not furnished in the fan package. An angle hanger allows the fan to hang level although the box sits at an angle. The fan can be supported by a hook, but you will need a block of wood in which to drive the hook. We recommend a short length of 4x4. Predrill it and bolt it to the beam or ceiling. The power wire can be stapled along the top or bottom edge of a beam and then connected into the electrical ceiling box. You can paint or stain the cable so it matches the beam.

4 Assembling Down Rod. Position the down rod through the canopy of the fan, using the diagram on page 65 to help you.

Run the electrical wires from the fan through the down-rod assembly. Use a stout string to pull them through if necessary. There will be power circuit wires and a switch wire. There also may be a fourth, grounding wire, depending on the brand and type of fan you are installing. The usual insulation colors identify the power circuit wires: black for the hot line and white. The switch wire is usually blue-insulated, but may have red insulation in some fans. The grounding wire will be bare or have green insulation.

Insert the down rod into the motor adaptor and fasten it. Insert the bolt provided. Then insert and spread the cotter pin. Finally, tighten the setscrew counter-clockwise .

5 Lifting Fan Into Position. With a helper, lift the fan into position at the ceiling without the blades attached. Put the vaulted ceiling mount or swivel into the hanger bracket. The fan now should be supported on the ceiling with a swivel-type ball-and-socket bracket or J hook.

6 Connecting Wires. Connect the fan to electrical power as shown on page 68. In a general hookup, the white wire goes to white, and black goes to black. The ground wire connects to the ceiling box via a screw or clip, or the ground is spliced to an incoming ground wire.

7 Hooking Up the Light Kit. If the fan has a light kit, remove the switch housing on the fan and the center screw. Screw the light kit onto the bottom plate. You will need a helper to hold the kit assembly in position while you connect the fan's blue wire to the light's black wire and the fan's white wire to the light's white wire. Use wire connectors for splices and wrap the wire connectors with electrical tape. Screw on the bottom plate and light, and attach the glass.

8 Assembling Fan Blades. Put the fan blades into the holder. Then fasten the holder to the fan motor. The unit is now assembled.

9 Testing Fan Operation. Turn on the power and make sure that the fan operates at all speeds, forward, and reverse. If the fan has a light kit, turn on the lights. If you notice any blade wobble, measure from the ceiling to the tip of each blade. If the blades are not even, exert light pressure on the blade that is out of alignment so it matches the other blades. You may have to loosen setscrews that hold the blades in position, if the fan has them. All blades must be uniform. Otherwise, the blades will wobble and jump and exert extra wear and tear on the fan mounting assembly.

Troubleshooting

If the fan does not run or the lights don't work, you can make these checks:

■ Is the power turned on at the main service panel? Has a fuse blown or breaker tripped while you were connecting the fan?

■ Is the wall or fan switch in the proper mode?

■ Is the light switch on the fan in the proper mode?

■ Are the power wires properly connected in the ceiling box?

■ Are the switch wires properly connected in the ceiling box?

Fan Switch on Fan

If the fan has a three-speed pull chain near the bottom of the motor, the blue and black fan wires are connected to the black-power wire in the box. The white-insulated wires are joined; the green ground wire is fastened to the ceiling box. Use wire connectors and tape. Optional light is operated independently of the fan by a pull chain.

Light Switch on Wall

The light is controlled by a wall switch. You control the fan with the three-speed chain. You also can have a separate wall switch to control a ceiling-fan light kit. The black wire in the fan connects to the incoming black power wire and to another black wire going to the switch. The blue switch wire connects to the blue fan wire. The white wire in the fan connects to the white wire in the box. The green wire connects to the green wire in the box or to the box itself.

Fan & Light Switch on Wall

In a dual-control wall switch, the black power wire is connected to a black wire going to the switch. The black and blue fan wires also connect to the switch. Connect the white wires; connect the green wires. If the box does not have a green grounding wire, connect the green fan wire to the box. Put pull chain on fan on high speed.

Transformer & Switch on Wall

If the fan has a three-speed transformer, you can control it with a wall switch or variable control. Connect the fan's black wire to the black switch wire. The blue wire is connected to the incoming black power wire and connects to the fan's blue wire. The white wires are connected and the green wires are connected. If you want variable speed on the fan, leave the pull chain on the fan at the high-speed setting.

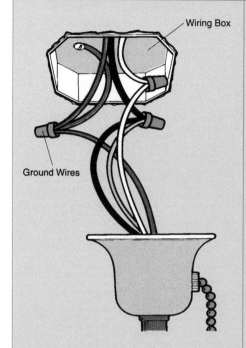

Fan Switch on Fan. Wiring hookup is for a three-speed pull chain. It provides for optional light kit, which is operated independently of the fan by a pull-chain switch.

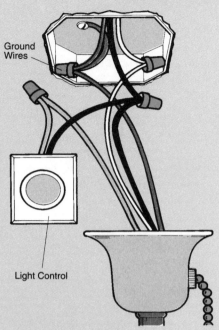

Light Switch on Wall. The light is controlled from a wall switch in this arrangement. You control the fan speed with its three-speed control, and the light from the separate wall switch.

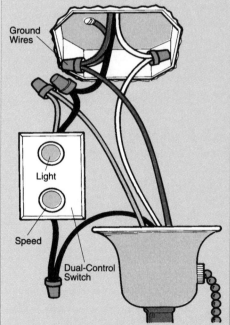

Fan & Light Switch on Wall. To control the fan and light from a wall switch, use a dual-control switch. Put pull chain on fan on high speed, although it may be on any speed.

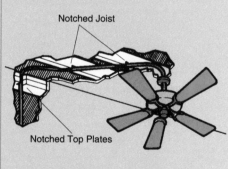

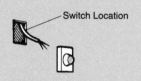

Transformer & Switch on Wall. For a supplemental variable control with a standard wall switch and three-speed transformer, use this hookup. Leave pull chain on high speed.

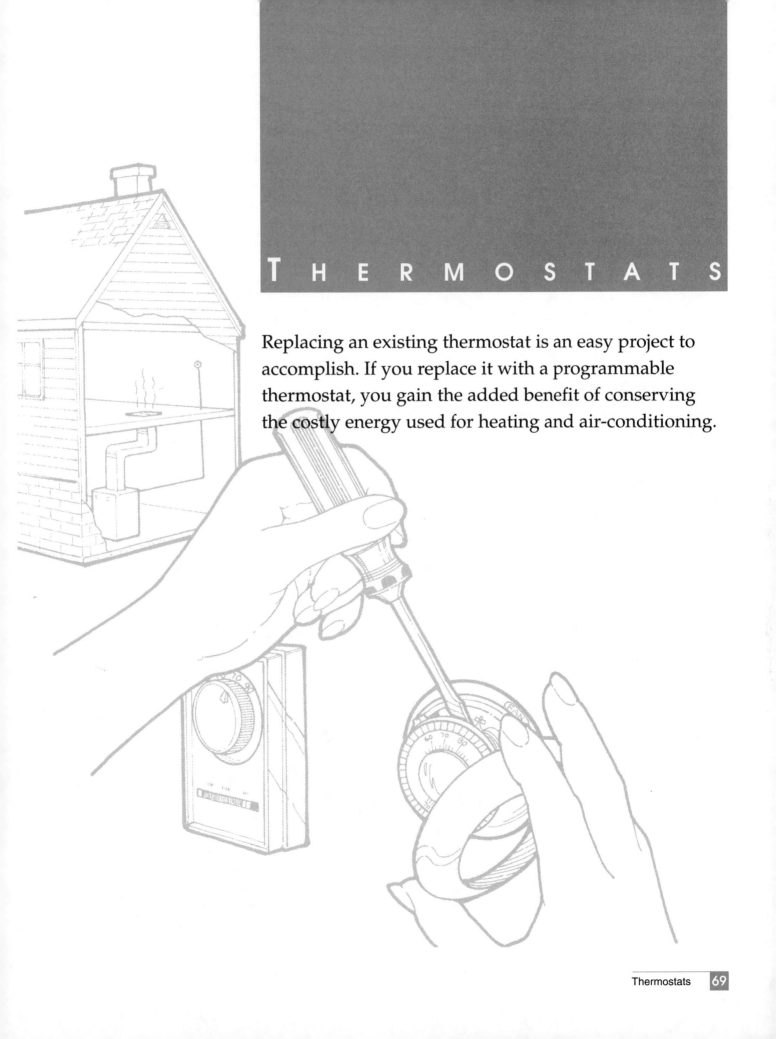

THERMOSTATS

Replacing an existing thermostat is an easy project to accomplish. If you replace it with a programmable thermostat, you gain the added benefit of conserving the costly energy used for heating and air-conditioning.

Replacing the Thermostat

If the existing thermostat in your home is malfunctioning and you want to exchange it for a new one of the same type and perhaps, the same brand, it is an easy project to do. Simply hook up the new unit to the same wiring.

If, however, you want to move the location of the thermostat, you will have to extend the wiring from the furnace. This involves fishing the wires to the location.

1 Turning Off Power. At the main service panel, turn off the circuit supplying the power to the furnace.

2 Removing Thermostat. Remove the thermostat cover, then remove the wires connected to the thermostat terminals. Remove the thermostat base from the wall.

3 Disconnecting Wires. When the old base is off, you will see two wires protruding from the wall. These run to the furnace. Mount the new thermostat base over these wires and pull them out of the wall about 3 inches. Drive in a couple of mounting screws, but do not tighten them. With a plumb bob or small torpedo-style level, level the base on the wall. Then tighten the screws. The base must be level or the thermostat will not work properly.

4 Connecting New Thermostat. Note the terminal connections on the mounting base. If your system has just two wires, you can connect either wire to either terminal screw. If the system has three wires, hook the wire with the white insulation to one terminal and the other two wires to the second terminal. If the base has three terminals and there are three wires, match the wires to the color codes on the terminals. Complete the job by installing the cover on the mounting base.

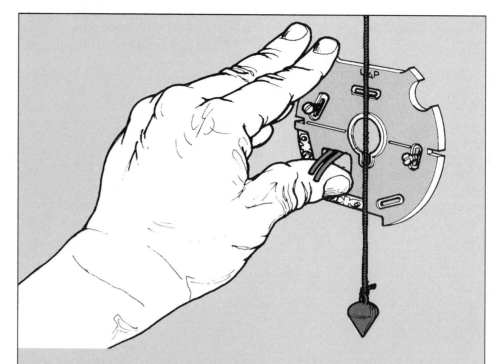

Mounting the New Base for the Thermostat. After you remove the old base, the furnace wires should be pulled through. Then level the new base on the wall and fasten it tightly in position. The base must be level on the wall so the thermostat will operate properly. Do not use the old base for the new thermostat.

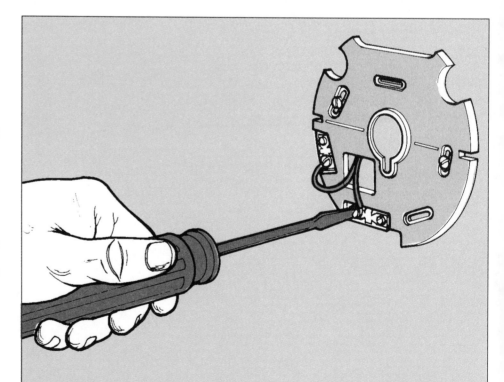

Mounting the Connections from the Furnace. Hook the furnace wires around the terminals in the direction that the screws turn (clockwise). You can hook either wire to either terminal. If there are three wires, hook the white wire to one terminal, and both other wires to the second terminal.

Adding a Programmable Thermostat

Thermostat manufacturers keep finding new and better ways to conserve the energy that serves heating and air-conditioning equipment. Perhaps the biggest and most efficient innovation now on the market is the programmable thermostat. It can be set to adjust the temperature in your home automatically for waking and sleeping hours. The thermostat also may include indicators that set different temperature levels for those hours when the house is usually occupied and for those hours when it is not.

Types of Units

Thermostats come in many different styles. Some units, called heat-only thermostats, control only the furnace. Other units control both heating and cooling, provided that the air-conditioning unit is mounted in your forced air furnace.

Since energy-saving, programmable thermostats vary considerably from one brand to another, be sure to get full installation and programming instructions with the unit when you buy it. For example, some thermostats require a complete week-long run-through to set the day and night temperature sequence how you would like it. Others operate on pins that simply may be set. If you misplace the instructions, do not try to improvise. Go back to the store or the manufacturer for the complete directions.

The Components

All models come with a wall plate that is fastened in place. The existing thermostat wires run through the plate. A subbase is attached over the existing wall plate and the programmable thermostat fits over this subbase. The contacts match exactly to make the hookup.

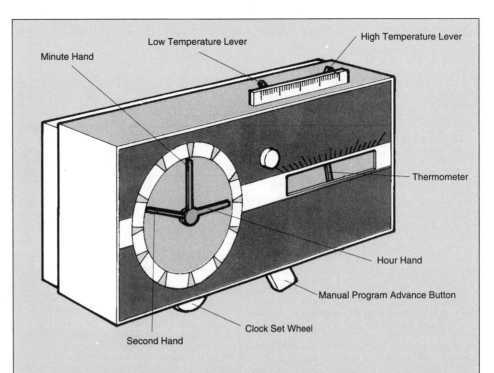

Programmed Thermostat Adjusts Temperature. With a programmed thermostat, you can regulate the temperature in your home according to your schedule and lifestyle. The thermostats are an easy, do-it-yourself connection, and they cost little more than a regular heating-cooling thermostat.

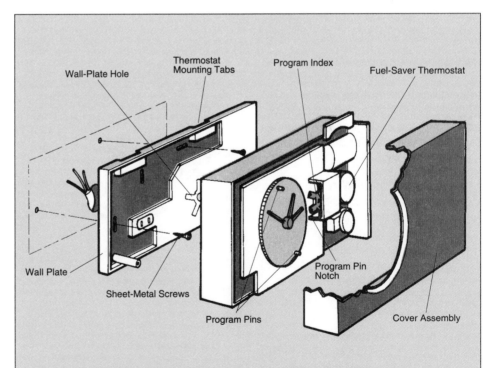

Programmed Thermostat Hooks to Existing Wiring. The wires from the heating-cooling unit are pulled through a wall base plate. Then the programmed thermostat is mounted to the base plate. Some programmed units use pins for heating-cooling settings. Others operate on a push-button system that sets the time and temperature.

Installing Programmable Thermostats

The thermostat shown in the exploded diagram on page 71 has a built-in clock that, coupled with programming, governs the heating and cooling cycle. The clock is run by a small battery that must be charged before the thermostat can function. From then on, the battery recharges while the thermostat is working.

1 Turning Off Power. At the main service panel, turn off the circuit supplying the power to the furnace.

2 Removing Old Thermostat. Remove the old thermostat from the wall. Lift off the cover plate and remove the wires from the terminals. Then remove the old base. Pull the wires a short way out of the wall. You may want to tag the wires so you can match them to comparable identification marks on the new thermostat. Plug the hole through which the wires protrude with loose fiberglass insulation. A warm or cold draft through the opening could affect the thermostat.

3 Connecting the Wall Plate. Connect the wall plate for the new thermostat. Pull the wires through the opening in the wall base and screw the plate loosely in place. Then level the plate and tighten the screws (see page 70).

4 Connecting the Wires. Connect the wires to the terminals of the new unit. The white wire goes to the terminal marked W. The red wire goes to the terminal marked R. If the wire is black, connect it to the W terminal. If there are three wires, connect the white and black wires to the W terminal.

If your system has a single transformer for heating and cooling, strip enough insulation from the wire to connect the wire to terminals RC and RH. This connection will be so noted in the manufacturer's instruction package.

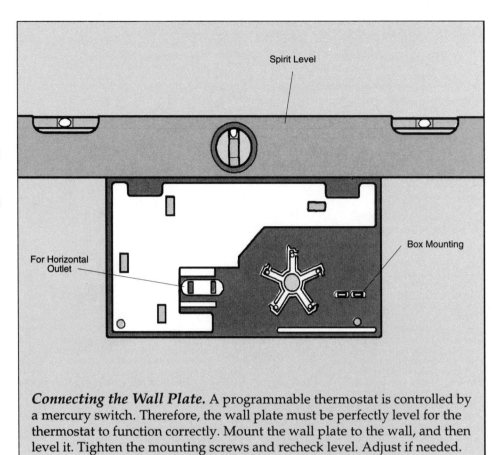

Connecting the Wall Plate. A programmable thermostat is controlled by a mercury switch. Therefore, the wall plate must be perfectly level for the thermostat to function correctly. Mount the wall plate to the wall, and then level it. Tighten the mounting screws and recheck level. Adjust if needed.

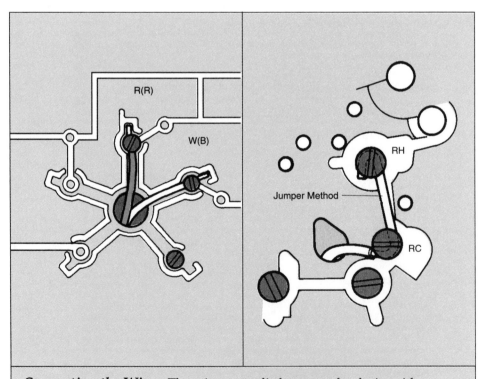

Connecting the Wires. The wires must lie between the design ridges. This prevents interference with the thermostat mechanism (left). If your system has a single transformer, strip enough insulation from the wire to connect terminal RC with terminal RH (right).

OUTDOOR WIRING

There are many good reasons to install outdoor wiring. Lighting is probably the primary reason, with power for appliances running a close second.

Outdoor Electrical Projects

Before starting an outside electrical project, contact the municipal building inspector to determine the requirements concerning outdoor wiring in your community. Some communities, require that a professional electrician make the final hookups. In other areas, the work must be inspected before it can be put into operation. Determine whether the municipal electrical code permits the use of Type UF cable, or if it specifies Type TW wire and conduit. Generally, local codes require that outdoor wiring be protected by conduit whenever it is installed aboveground. If the wiring is to be buried, most codes allow Type UF cable. Some require that Type TW wire and conduit be used.

Caution: *Always turn off the power at the main electrical service panel before working on a circuit.*

Outdoor Receptacles

Outdoor light fixture boxes possess the same characteristics as boxes for outdoor receptacles. To inhibit moisture, there is a thick gasket that seals the joint between the fixture and the fixture box cover plate.

Electrically, the fixture is set up the same as an indoor fixture. Conductors connect the fixture to wires coming from the power source. Ground wire connections also are the same.

Note: For outdoor fixtures use weatherproof bulbs that resist shattering in severe weather.

Outdoor Lighting

The receptacles used outdoors are the same type, size and shape as the ones used indoors. However, outdoor receptacle boxes are made of heavy metal. Instead of knockout plugs, outdoor receptacle boxes have threaded openings into which

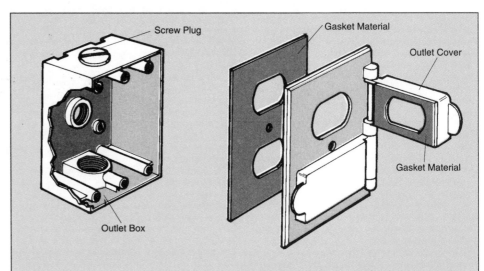

Outdoor Receptacles. These receptacles are made of extra-thick metal with screw fittings and gaskets between faceplates and openings to the outlets. They may be found in the lawn and garden department of stores.

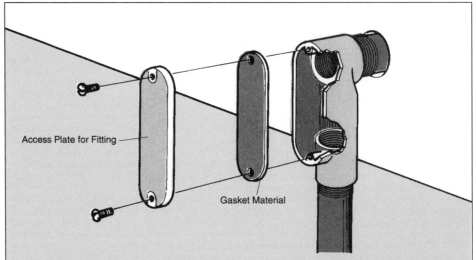

Outdoor Conduit. Fittings for conduit and wire used outdoors include these basic products. An LB fitting is L-shaped and has a back opening for conduit. The fittings have weatherproofing gaskets.

conduit is screwed. If conduit is not used, UF cable passes through a thick plug that is screwed into the threaded opening.

Cover plates for outdoor receptacles are made of heavy-gauge metal. They are outfitted with spring-action doors that close tightly over the outlets when they are not in use. The joint between the receptacle box and box cover plate is fitted with a thick gasket to prevent moisture from seeping into the box.

Outdoor Switches

Heavy cast metal is used for the construction of switch boxes that will be mounted outdoors. Cover plates for the boxes are made of the same material and are outfitted with weatherproof gaskets. An ordinary toggle type of switch is completely encased in the box. Even the toggle is covered. A lever on the outside of the cover extends into the box to engage the toggle. The lever is used to turn the switch on and off.

Outdoor Lighting Guidelines

Whether you plan to light the walkways and paths for safety purposes, or spotlight a lawn feature, it is a good idea to draw a plan of the wiring scheme you want. This will save you plenty of time and help you estimate the materials needed. Use the following guidelines to plan your installation.

■ Provide enough light to meet the needs, but do not overdo it. Too much light can ruin the atmosphere and can cause glare. Several strategically placed small lights are better than one large light.

■ Place the lighting fixtures so only the light is seen, not the source.

■ To achieve a visual effect in any area, backlight architectural or landscaping features. This creates shadow and depth.

■ Positioning a series of lights can extend the interior lights bordering a backyard. Exterior lighting should complement interior lighting.

■ Try to keep the distance between lights to a minimum. The more spread out the lighting design, the more costly the installation.

■ For safety, provide at least 5 lumens of light at all walkways, ramps and stairways.

■ Stairway lights should shine down toward the risers so that the treads will not be in shadow.

■ Place all fixtures and receptacles so that they are easily accessible, for future maintenance. Keep the total distance between fixtures to a minimum.

■ If you have to dig to bury cable, do not dig until you contact the utilities. They will need at least three days notice to plot underground wiring and plumbing for you. Driving a spade blade through an electrical cable can be dangerous.

Common Hook Ups

The procedures for hooking up a receptacle are the same as the procedures for hooking up outdoor lighting or other outdoor accessories.

Equipment

Outdoor electrical equipment, such as fixture boxes and receptacles, is especially manufactured to meet codes and resist the elements. Do not use electrical products specified for indoor-use outdoors.

Connectors & Fittings

An LB fitting (see page 74) is a right-angle connector that is used with a conduit to bring cable through the wall of a house. The fitting routes cable toward a trench that has been dug from the house to the area where the electricity is needed.

LB fittings are threaded on both ends. Conduit passing through the house wall to the outside is screwed to one end. Conduit leading down the side of the house to the trench is screwed to the other end. Thus, cable is enclosed in the metal to provide an efficient seal from the time it leaves the house to the time it enters the ground. LB connectors are outfitted with thick gaskets and metal cover plates.

Box extenders are used when tapping an existing outdoor receptacle or fixture junction box for power. The extender may have a nipple and a 90-degree elbow so that the wires may be brought from the fixture, through the conduit, to the point where power is needed.

Outdoor Conduit

Three types are available, but check the codes before you buy.

■ Rigid aluminum and rigid steel conduit provide equal protection to the wires that pass through them. Rigid aluminum is easier to work with, but if it is going to be buried in concrete, first coat it with bituminous paint to keep it from corroding.

■ Both types of metal conduit come with a variety of fittings, including elbows, offsets, bushings, couplings and connectors. If offsets and elbows do not provide the necessary turns in rigid metal or EMT conduit, you will need a bending tool called a hickey.

■ Nonmetallic conduit is made from either polyvinyl chloride (PVC), which is normally used aboveground, or high-density polyethylene, which is suitable for burial. If PVC is going to be exposed to direct sunlight, it must be labeled as suitable for use in sunlight. But before you purchase nonmetallic conduit, be sure you check local codes. Do not assume that it is approved. An inspection might require you to replace the nonmetallic materials with another product; this can be very costly and time-consuming.

Outdoor Accessories

Stores offer a wide variety of electrical accessories manufactured especially for outdoor usage.

Included are all types of lighting fixtures, such as spotlights and lampposts, as well as outdoor cooking appliances, pool lights and devices to control insects. Most of the accessories run on regular 120-volt housepower, so connecting power from your home to the device is easy enough for a do-it-yourselfer to do.

The job consists of tapping into the power inside of the house and running a cable through the wall to the exterior. Then the connections are made. The hardest part of the job is digging wire trenches.

Installing Outlets; Directly Into Interior

This project demonstrates how to install a single exterior outlet to provide power for appliances and tools.

Through the Siding

1 **Locating the Outlet.** If at all possible, locate the exterior outlet directly opposite an interior outlet. This way, you can use the same power source for both outlets. Shut off power to circuit that operates the outlet. Remove faceplate and outlet.

2 **Drilling the Hole.** With a long 3/4-inch drill bit, drill a hole through an opening in the back of the box through the sheathing and siding. Outside, locate the drilled hole. With a keyhole or saber saw, cut away the sheathing and siding to fit the exterior box that you will install. Set the saw so the blade will not enter the interior box.

3 **Connecting the Outlet.** Remove the back knockout on the cast metal box and screw this box into the house with knockout hole aligned with the hole in the wall. Insert 10 inches of cable in the hole. Connect the cable terminals of inside receptacle and the cable outside to new GFCI receptacle. Then install a waterproof gasket and faceplate over the outside outlet.

Through Masonry Walls

1 **Drilling the Hole.** Outline the shape of the box on the concrete block with masking tape. Within that border drill a series of holes using a power drill with masonry bit. Clean out the area with a cold chisel and baby sledge. Wear safety glasses and gloves. Drill a hole through the wall, matching a knockout in the back of the box. The cable will run through this hole to a junction box on interior wall.

2 **Connecting the Outlet.** Adjust the ears of the exterior box so the box will extend about 1/16 inch from the block. Cement the box in place. Connect outlet and install a weatherproof cover.

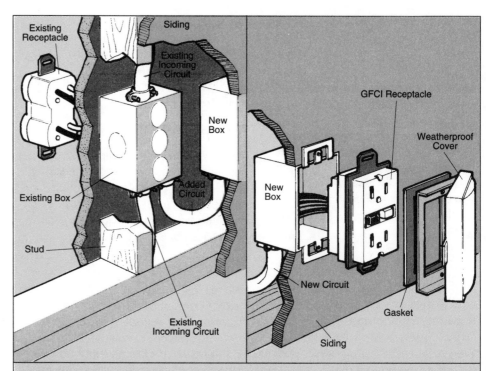

Through the Siding. Turn off power and remove existing outlet from interior box. Drill hole through back of box to exterior. You will tap this outlet (above left). Cut hole in siding and sheathing from outside and insert exterior box. Run cable through knockout to inside box and make power hookup (above right).

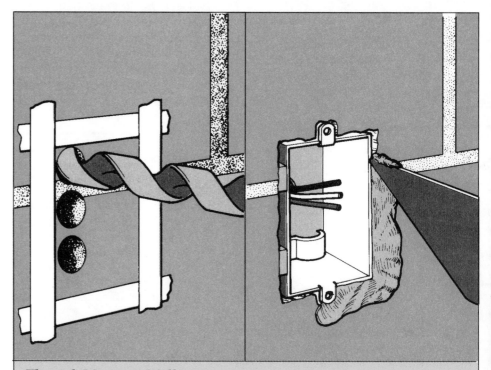

Through Masonry Walls. Drill series of holes in concrete block for exterior box (above left). Remove back knockout, insert box, and drill hole to interior for power cable. Mortar exterior box in place (above right). Connect outlet to power in interior junction box, which may be surface-mounted to block with masonry shields.

Installing Outlets; Using Conduit

There are two main ways to extend electricity from the house to the outside. You can run the power cable through the basement (or basement crawl space) or through the attic.

Through Basement

1 Locating Access to Cable. It may be near a water pipe that extends through the wall or at a corner. The spot where you go through the wall should be at least 3 inches from a joist, sill plate or floor to allow clearance for a junction box.

2 Identifying Exit Point. Outside, measure from the common reference point to the spot selected for the exit. If the spot is on the foundation, make sure the spot does not fall on a joint between concrete blocks or where two pieces of siding join. The spot has to provide a firm base for the LB fitting. At this point outside, drill a small hole through the wall to verify that the path is clear. If the hole is in a block wall, do not drill through the top block. Blocks below the top have a hollow center; top blocks often are filled with concrete.

3 Drilling the Hole. Use a star drill and baby sledgehammer to cut the opening for the extender in masonry. Wear safety glasses and gloves while working.

4 Mounting the Box. Back inside, open one of the knockouts from the back of a box and mount the box so the hole matches the hole through the wall. The box is mounted with masonry shields (anchors) and screws.

5 Digging the Trench. Outside, dig the cable trench. Be sure to check codes.

6 Attaching Conduit. Screw a nipple, long enough to extend from inside the box through the hole to the outside, onto an LB fitting screw. Outside, attach conduit to the LB fitting and run the conduit down the side of the house to the trench. Then seal the joint around the fitting with quality caulking compound. Inside, secure the nipple to the box with a connector. The opening is now ready for the cable.

Through Attic

1 Mounting the Box. Hold the assembly against the overhang of the roof so the box and nipple are against the soffit and the conduit is against the wall. Try to run the conduit near a downspout to make it inconspicuous.

2 Drilling the Soffit Hole. Mark the soffit where the cable will pass through the soffit into the box. Use a 1⅛-inch bit to drill a hole through the soffit for the cable. Then remove a knockout to correspond with the hole and fasten the box to the soffit with screws.

3 Attaching Conduit. Run the cable from the attic power source and out the hole in the soffit. Clamp the cable to the box. With conduit straps, strap the nipple and conduit into place and complete the installation by running the conduit down the side of the house. The path is now ready for the cable installation. Be sure to check the codes for the type of cable and/or conduit you can use.

The Final Connection

You will have to pull the cable through the pathway you have made for it. Then the cable can be connected to the power source inside the house. The best plan is to complete the entire project first— hooking up the outside lights and appliances—before connecting to power. If you are creating a new circuit, have a professional electrician connect the cable to the main service panel. If you are connecting to an existing outlet, make sure the circuit has enough amperage to handle your needs for outdoor power.

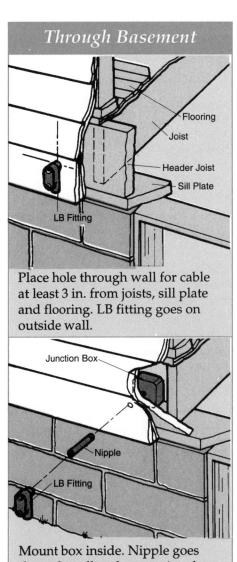

Through Basement

Place hole through wall for cable at least 3 in. from joists, sill plate and flooring. LB fitting goes on outside wall.

Mount box inside. Nipple goes through wall and screws into box. LB fitting screws onto nipple; conduit screws into fitting.

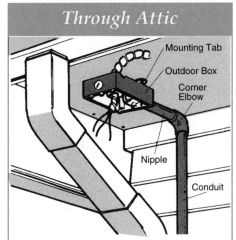

Through Attic

Cable runs from attic to box fastened to soffit exterior. Conduit drops to trench on ground. Run conduit next to downspout, if possible.

Installing Outlets; Away from the House

Running the cable from the house out to the yard, pool, garden or elsewhere involves digging a shallow trench and building an anchor for each receptacle.

Digging a Trench

First, call the utility company for a plot of pipes and wiring that may be running underground on your property. In some areas you are required by law to do this.

Check your local code on depth requirements. Generally, cable not in conduit (but with proper insulation, see page 16) must be buried at least 24 inches deep, with expansion loops as shown at far right.

Put intermediate metallic conduit at least 6 inches deep, rigid nonmetallic conduit at least 18 inches deep.

If the wires have to go under a walkway or driveway, dig the trench up to the obstruction. Then continue the trench on the other side of it. Cut a piece of conduit 10 inches longer than the span. Hammer a point on one end of the conduit. Now hammer the conduit under the obstruction. When it appears on the other side, cut off the point with a hacksaw. You can connect another piece of conduit to it or run the cable through the conduit under the obstruction.

Installing Receptacles

The receptacle should be at least 12 inches above ground level and anchored underground. Do this by laying the cable or conduit either through the center opening of a concrete block or through a coffee can filled with concrete. Since so little cement is required for this job, we suggest that you buy already-mixed concrete, usually sold in 80-pound bags.

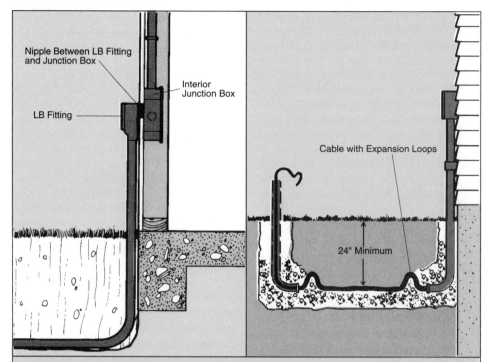

Digging a Trench. The drawing left shows the hookup for outdoor wiring as it leaves the house on its way through a trench to a receptacle or appliance in the yard area. The cable should be buried at least 2-ft. underground, or according to local code. Loop the cable as shown right, to allow for expansion.

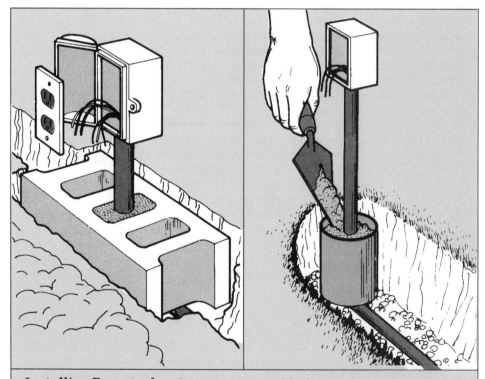

Installing Receptacles. Lower a concrete block over the positioned receptacle and conduit/cable for an anchor. Fill space with concrete (left). You can use a large tin can as a receptacle anchor. Cut both ends open and lower the can over the box and into the ground. Fill with concrete (right).

AC Alternating current. The type of current found in most home electrical systems in the United States.

AWG American Wire Gauge. A system of sizing wire.

Ampere, amperage, amps A unit of measurement that describes the rate of electrical flow (current). Amperes are measured in terms of the number of electrons flowing through a given point in a conductor in one second when the electrons are under a pressure of one volt and the conductor has a resistance of one ohm. Amps multiplied by volts equals the number of watts available for use (VxA=W). Cables and wires are rated by their amperage—the number of amps they can safely handle.

BX cable Electrical cable wrapped in a protective, flexible, metal sheathing. BX contains at least two conductors.

Ballast Device that controls the current in a fluorescent light.

Black wire In a cable, the wire that functions as a hot wire.

CSA Canadian Standards Association. See National Electrical Code.

CO/ALR Marking that designates switches and receptacles that may be used with aluminum wiring.

Cable Two or more wires grouped together inside a protective sheathing of plastic or metal.

Cartridge fuse Cylindrical fuses that carry higher voltage than a plug fuse.

Circuit breaker A protective toggle switch that automatically switches off (trips) the power to a circuit in the event of short circuit or overload.

Common The brass-colored hot terminal on switches and outlets. Also the black-insulated hot wire.

Conduit Metal or plastic tubing designed to enclose electrical wires.

Continuity tester A device that tests the ability of a circuit to sustain the flow of electricity. Never used with the power turned on or to test if the power is on in a circuit.

CU/AL Marking that designates receptacles and switches that may be used with copper or copper-clad aluminum wire.

DC Direct current. The electrical current supplied by a battery and often an engine-driven generator.

Double-pole switch A switch with four terminals that controls a single major appliance. The toggle is marked ON/OFF.

End-of-the-run Box with its outlet or switch at the final position in a circuit. Only one cable enters the box.

Energy efficiency rating (EER) The relative amount of energy consumed by an appliance. The higher the EER, the more efficient it is.

Four-way switch One of three switches controlling one outlet or fixture. The other two switches are three-way switches. A four-way switch has four terminal screws. The toggle is not marked ON/OFF.

Fuse A safety device designed to protect house circuits. A metal wire inside the fuse melts or disintegrates in case of overload or short circuit, thus shutting off the current. See also Cartridge fuse.

Green wire In a cable, the wire that functions as a ground wire.

Ground Fault Circuit Interrupter (GFCI) A safety circuit breaker that compares the amount of current entering a receptacle with the amount leaving. If there is a discrepancy of .005 volt, the GFCI breaks the circuit in 1/40 of a second. Usually required by code in areas that are subject to dampness.

Ground wire Wire that carries current to earth in the event of a short circuit. The ground wire is essential to the safety of your house wiring system and of its users.

Incoming wire Hot wire that feeds power into a box.

Jumper wires Short lengths of single wire used to complete circuit connections.

Junction box Metal box inside which all standard wire splices and wiring connections must be made.

Kilowatt This measurement equals 1,000 watts; (kw/h) equals 1,000 watts used for one hour.

Middle-of-the-run Box with its outlets or switch lying between the power source and another box.

NM cable Cable for use only in dry locations.

NMC cable Cable used in damp and dry locations, but not in wet locations.

National Electrical Code Body of regulations spelling out safe, functional electrical procedures. Local codes add to but not delete from NEC regulations.

Overload Too great a demand for power made on a circuit.

Pigtail A short piece of wire used to complete a circuit inside a box.

Polarized plug Plug with one prong larger than the other. It is a safety feature against electrical shock.

Recoded wire White-insulated wire that has been taped black. The recoding indicates that wire now carries power.

Red wire In a cable, the wire designated as a hot wire. Usually used as a switch wire in three-way switches.

Romex Plastic-sheathed cable containing at least two conductors.

Service panel The point at which electricity enters a house wiring system.

Short circuit A fault that occurs when a bare wire touches another bare wire carrying electricity, or touches any metal part of an appliance.

Starter A switch in a fluorescent light that closes the circuit only when sufficient power is available.

Switch loop Installation in which a ceiling fixture is installed between a power source and a switch. The power bypasses through the fixture box to the switch. The switch then sends power to the fixture itself.

TW wire Type of wire most often used in home circuits, raceway and exterior conduit.

Three-way switch One of two switches controlling a single outlet or fixture; it has three terminals.

Traveler wire Transfers electricity from one three-way switch to another.

UF cable Cable for use in wet, outdoor locations; also used inside.

Underwriters' Laboratories Independent organization that acts as a watchdog over electrical equipment. The UL label is granted only to equipment that meets minimum performance standards.

Volt, voltage Unit of measurement of the electromotive force of a current. One volt equals the amount of pressure required to move one ampere through a wire that has a resistance of one ohm. Volts multiplied by amps give the wattage available in a circuit (VxA=W).

Watt, wattage Unit of measurement of the amount of electrical power required or consumed by a fixture or appliance. See also amps and volts.

White wire Wire with white insulation that functions as a neutral wire.

Wire connector Plastic cover for a wire splice. The inside is threaded metal.

Zip cord Cord designed to easily split when pulled down the middle.